THE ENERGY TECHNIQUE

THE
ENERGY
technique

Simple secrets for a lifetime of
vitality and energy

VERA PEIFFER

Thorsons
An Imprint of HarperCollins*Publishers*

Thorsons
An Imprint of HarperCollins*Publishers*
77–85 Fulham Palace Road
Hammersmith, London w6 8jb

Published by Thorsons 1999

10 9 8 7 6 5 4 3 2 1

© Vera Peiffer 1995

Vera Peiffer asserts the moral right to
be identified as the author of this work

A catalogue record for this book
is available from the British Library

ISBN 0 7225 3792 1

Printed and bound in Great Britain by
Woolnough Bookbinding Limited,
Irthlingborough, Northamptonshire

FOR DUNCAN

MY SPECIALEST PERSON

Contents

Introduction

If you want to see energy in motion, the best place to find it is a children's playground. Sit down on a bench and spend an hour or so observing those five-to-ten year-olds climbing and running and jumping, at the same time laughing and shouting at the top of their voices.

What would it be like to have this vitality and physical strength? Imagine how life would change for the better if you could translate this energy into your adult life – waking up in the morning feeling refreshed and wide awake, then eating a small breakfast very slowly, then bouncing off to work, eager and ready for the excitements of the day, working happily and whole-heartedly and coming back home, still full of beans, going off to the theatre or to see some friends, and then falling into bed and into a deep and undisturbed sleep.

This may sound too good to be true, and to a certain extent it is. As adults we can never be quite as energetic as we used to be as children, partly because of our altered circumstances, such as having to go to work and taking on board a lot of responsibilities to do with our job and our family. The good news is that we can nevertheless do a lot to increase our energy levels substantially, and this is what this book is about.

So roll up your sleeves, take the brake blocks away from under your feet and start rolling. If others can do it, so can you!

PART I

The Concept of Energizing

WHAT IS ENERGY?

In our everyday language, there are many ways in which we can express that we are feeling energetic – we feel 'full of beans', we are 'raring to go', 'bouncy', 'bright-eyed and bushy-tailed', 'motivated', 'buzzing' or 'on top form'. When we feel a lack of energy we speak of being 'tired', 'exhausted', 'weary', 'wiped out', 'down', 'uninspired' or 'sluggish'.

You will notice that we use the word 'energy' to describe two different aspects. On the one hand, we refer to the state of our bodies and how this state varies between fully charged or drained of energy; on the other hand, we refer to our psychological state of energy which can differ from being up and positive to being down, disheartened and vulnerable.

We have all experienced how our energy levels concerning one aspect of our person will have a direct knock-on effect on the other aspect. If we overwork continuously, we do not just run down the physical batteries but we also influence our emotions negatively. Depending on our personality, we get more irritable or more tearful the more tired we get. Equally, if we are under a lot of psychological strain we are much more likely to fall ill, and illness is always a warning sign that we have an energy crisis in our bodies.

The whole process of creating energy starts with the sun. Together with carbon dioxide gas from the air and water from the soil, sunlight is the energy source which forms nourishment for plants. The process whereby sunlight is absorbed by the plant's chlorophyll is called *photosynthesis*. During photosynthesis, sunlight energy is used to combine carbon dioxide with water. This produces glucose sugar, and oxygen gas is released into the air as a waste product. The glucose is now used to make all the carbohydrates, fats and proteins which make up the plant body.

Plants are eaten by animals (herbivores), and these animals are in turn eaten by meat-eating animals (carnivores), both herbivores and carnivores are in turn eaten by us humans. In this food chain energy is passed from the plants to the herbivores and then on to the carnivores/us, and in this passing-on process a certain amount of the original energy is lost.

When we eat food, our bodies break down the ingested food into carbon dioxide and water. This process is called internal respiration and means that the energy contained in the food is now being released into our system. In order to make energy release possible, oxygen is needed, which we take into our system by breathing.

At this stage, the energy released cannot as yet be used by the body. First, the energy needs to be converted into molecules of adenosine triphosphate (ATP). ATP stores energy temporarily until it is required for metabolism, such as use of muscles for example; in this respect you could compare ATP to loose change in your pocket – you carry it around with you until you need to spend it.

We need to eat to live because eating supplies us with energy. We have to eat enough to have sufficient energy to be able to function and work throughout the day. If we do not eat enough over a long period of time, we have no energy reserves, we feel tired easily and are more prone to illnesses. If we eat too much, the excess food is converted into fat which puts strain on the heart and on the joints, all of which have to work harder because of the excess weight, and this in turn makes us feel tired. What constitutes the right amount of food will of course depend on a variety of factors, such as age, sex, body size and occupation.

But even though eating food helps us gain energy, energy is also needed to digest the food; this is why you feel tired after a meal, especially when you have had several courses or if you have eaten foods that are hard to digest, such as meat. In fact, there is no other function of the human body that demands as much energy as your digestion; not even running or cycling needs as much! This means that if you eat foods which are difficult to break down, your body will use up a lot of energy that would otherwise be available for activities other than digestion.

As mentioned, the body needs oxygen to break down foods. It is therefore important that we ensure that enough oxygen enters our bloodstream to help with the process. This is where exercise plays an important role. Not only does it increase the efficiency of the heart muscles and the muscles used for breathing, but it also helps digestion and increases stamina. If you exercise regularly, you will find that your energy levels improve and you can generally be more active without getting tired.

So this is the physical side – but what about the emotional side? How can we create a 'buzz' in our life so that we feel motivated and mentally awake? Where do these 'peak experiences' come from, which humanist psychologist Abraham Maslow (1908–70) describes as being moments in life where you feel full of joy and energy because you are at one with yourself and everything around you?

Apart from being healthy, mental and emotional energy can be derived from several sources. One is to spend your days doing things you enjoy and which are meaningful to you. If you are stuck in a job which you dislike it will drain your energies faster than pursuing a career that is interesting and rewarding. If you feel burdened by having to look after your children you are more likely to get upset and irritable, and these negative emotions drain your energy, whereas if you enjoy being with your children you are more balanced and will not exhaust your energy store. But even if you do not have a job that is totally 'you', there is an opportunity to make up for it by choosing a pastime which you can really enjoy. Boredom and lack of gentle challenges are great energy-stealers. That is why it makes sense to cultivate hobbies *before* you retire, so that once work is no longer the main focus of your life you have built up a range of entertaining and fulfilling activities that keep you mentally and physically fit.

Another factor in life that will increase your energy levels is that of having relationships with positive people, be they family, friends or a partner. Spending time with people you like and who like you is one of the most energizing activities!

In the mental sense, energy is whatever puts a sparkle in your eye – so I am afraid, folks, that watching television is out, because I have certainly never seen anyone who jumps up afterwards and is ready to burst into action …

WHAT IS YOUR ENERGY QUOTIENT?

As promised, here is a questionnaire that will allow you to assess how high or low your energy levels are present. Please take a bit of time over answering the questions and think carefully whether your answer should be 'always', 'often', 'sometimes', or 'never'. In answering, take into consideration the last two or three months and think of both your private and your work life.

For every time you answer 'always' to a question, give yourself 3 points; for every time you answer 'often', mark down 2 points; for every time you answer 'sometimes', mark 1 point, and give yourself 0 points for each 'never'.

1. Do you find it difficult to concentrate at work/when reading/when watching TV?

2. Does any physical exercise leave you exhausted for days on end?

3. Do you feel tired even when you have had sufficient sleep?

4. Do you find it difficult to get interested in anything?

5. Does any energy you have built up during your holidays vanish on your first day back at work?

6. Do you feel low and listless?

7. Do you suffer from a lack of coordination which makes you clumsy?

8. Do you suffer with poor short-term memory?

9. Are you out of breath when you have walked up a couple of flights of stairs?

10. Do you stammer or stutter?

11. Do you suffer from twitches or facial ticks?

12. Do you want others just to leave you alone, no matter whether they are friend or foe?

13. Do you stumble easily?

14. Are you unable to pursue any activities (excluding watching TV) once your day's work is done, simply because you are too tired?

15. Do you get cold hands and feet?

16. Do you find it difficult to fall asleep or stay asleep, or do you wake up far too early in the morning?

17. Does your back get tired easily?

18. Do you get colds or sore throats?

19. Do you cry easily over nothing?

20. Do you feel dizzy?

21. Do you sweat during the day or at night?

22. Do you suffer from heavy periods?

RESULTS

50–66 points:
Please go and speak to your GP or a specialist. There is a chance that you suffer from severe Candida or possibly ME (Myalgic Encephalomyelitis), also known as CFS (Chronic Fatigue Syndrome), especially if you find that your lack of energy is made worse by exercise. If so, there are special exercises you can do (for only half a minute three times a day – *see Part Two, Exercising chapter*). Stick to very gentle activities only. Concentrate on the sections on Food, Resting, Music and Visualization later in this book.

35–49 points:
Your energy levels need urgent attention, especially if you are close to the 50-points mark. You will have to work at

both the nutritional side and the building up of physical stamina and mental resilience. Start gently and build up slowly.

20–34 points:
You are in quite good shape, especially if you are in the lower 20-point range. You can afford to please yourself in choosing from the energy 'menus' in this book. Even if you just work on one particular area from the Resting section (*Part Three*) or from the Interacting and Exploring section (*Part Four*), for example, you should see improvements soon.

0–19 points:
Why don't you give this book to someone who really needs it . . . ?

Should you have scored in the higher numbers, please do not panic. You can still use a lot of the self-help exercises in this book to help your body to recover its strength. It will take you longer than others to see results, but as you build up strength at a slow rate, you can be sure that you will prevent relapses. It is worth taking your time over it. No matter whether you scored 10 or 66 points, let us have a look at what causes can lie at the bottom of a loss of energy. You may well recognize some issues as being pertinent to your own situation, and this will help you tackle them more efficiently.

HOW ENERGY GETS LOST

When we finally realize that we feel run down and tired a lot of the time, we are usually well into the situation that caused our lack of energy in the first place. The original reason for our fatigue might have been a sudden one – like some emotional shock, an accident or other acute stress – but more often than not we find that it was a gradual process. We somehow lost our energy on the way, and it can sometimes be quite difficult to pinpoint when this loss began. It is only when the discomfort gets too great and our normal 'action radius' is infringed upon that we sit up and take notice.

There are a variety of psychological, physical and environmental reasons that can underlie depleted energy levels, and it is often the case that a combination of several causes act together. Especially with environmental factors, it is not always possible to eliminate the causes altogether, but as long as we are aware of them we can do our best to avoid them.

THE IMMUNE SYSTEM

Our bodies are well equipped to fight invading bacteria, viruses, fungi and animal parasites. When you have hurt yourself and the wound becomes infected, the cells in the injured area release histamine and other chemicals which make it possible for plasma proteins to come to the rescue against the invading bacteria. At the same time, white blood cells called 'scavengers' are mobilized which swallow up the bacteria. Once the bacteria are caught inside the scavenger cell, they are destroyed by digestive enzymes. Should the infection be particularly persistent, even stronger white cells called monocytes are activated, and if even more support is needed, the lymphatic system jumps into gear.

The lymphatic system has two main tasks: it drains the body of fluid that sits between cells, and it also picks up dead material and foreign bodies. The lymphatic liquid is moved around the body indirectly through the blood circulation, which itself is driven by muscle activity, for example by exercise. It is in the lymph nodes – the large lymphatic vessels located in the neck, armpits and in the groin – that white lymphocyte cells are produced. The two main ones are the B-cells, produced in the bone marrow, and the T-cells, produced in the thymus gland which lies beneath the breast bone.

B-cells produce antibodies which fight bacteria and viruses. They are supported in this role by T-helper cells, a particular type of T-cell, which stimulate the production of scavenger cells. T-cells, as opposed to B-cells, do not make antibodies. Instead, so-called T-killer cells attack infected cells directly by bombarding them with cell poison. Once the invading bacteria or virus have been defeated, T-suppressor cells ensure that the immune system is switched off again and no more antibodies are produced by the B-cells.

Together with the liver, the spleen and the kidneys, all these white cells represent a formidable defence army that works day and night to keep us healthy. Naturally, anything that reduces the number of white cells will weaken the body's ability to combat bacterial or viral infections, and the first warning sign that we get of this is that we feel tired a lot of the time . . .

ENVIRONMENTAL FACTORS

So what is it that can weaken our immune system so that our energy levels drop and we become easy prey to illnesses?

Unfortunately, quite a few noxious substances are found nowadays in our environment. The air we breathe is contaminated by fumes and gases from car exhausts and factories. Carbon monoxide, nitrogen oxides and hydrocarbon all put stress on the immune system by reducing the growth of B- and T-cells and by damaging the mast cells in our respiratory passages, with the result that antibodies are no longer produced. The result is asthma, bronchitis and lung cancer.

As our bodies are already bombarded with environmental fumes and gases from dry-cleaning, glue, paints, insecticide and pesticides, the least we can do is stop adding to the toxic burden by making every attempt to stop smoking. Once a non-smoker, you will find that your increase in energy is quite spectacular!

We also have to be careful about the quality of our drinking water. As more and more nitrates seep into the ground through sewage and the application of fertilizers, it makes sense to use bottled water or to install a water-filter system in the kitchen which allows us to use good-quality water not just for drinking but also for washing and cooking vegetables in and for making soups and stock with.

The reason for this is that the immune system can only work properly if the heart is in good shape. Nitrates however, combine with haemoglobin to prevent oxygen from being transported around the body in sufficient quantities and, as we all know, a lack of oxygen makes us feel sluggish and listless. And as if this were not enough, nitrates also form an unhealthy allegiance with bacteria in the digestive tract, which can ultimately result in cancer, and malignant cells flourish in an environment low in oxygen!

As we are looking at environmental influences, it is also worth mentioning atmospheric circumstances which can add to you feeling tired. This can occur in an environment where the atmosphere is too high or too low in its humidity content. In many northern countries we tend to have a problem with a humidity level of between 45 and 55 per cent, not because of the weather but because of the fact that we tend to insulate our centrally-heated homes and offices by sealing off windows and doors with double glazing to help keep the heat in. This artificial environment can lower humidity down to a figure below

10 per cent, and this results in you feeling worn out. Professionally installed humidifiers can alleviate this problem.

Another atmospheric problem that affects our sense of well-being is that of ionization. Air that is charged with negative ions has an invigorating effect on the body, as it stimulates the oxygenation of the blood. The uplifting effects of negative ionization can best be felt by the seaside. On the other hand, an environment which contains synthetic materials, for example those found in carpets, curtains and upholstered furniture, will create an atmosphere of positive ions, especially if the natural exchange of air is prevented through double glazing.

Lastly, a word should be said about the importance of full-spectrum natural daylight. It appears that bright artificial light suppresses the production of a hormone called melatonin in humans. This hormone is linked with seasonal rhythms in animals and is responsible for some animals going into hibernation. Most of us will have experienced the positive effects of the first rays of spring sunshine – the lighter, longer days give us a lift. Although most of us are not unduly influenced by the darker months between October and April, some people are severely affected. They suffer from a condition called Seasonal Affective Disorder (SAD). The condition seems to be caused by a biochemical imbalance in the hypothalamus due to the shortening of daylight hours and the lack of sunlight in winter. The symptoms of SAD include a desire to oversleep and difficulties in staying awake and concentrating. In severe cases, the sufferer feels miserable, hopeless, depressed and severely lethargic.

Treatment that has proved effective in a great number of diagnosed cases is light therapy. This consists of exposure for up to four hours a day (average one to two hours) to very bright light of a specially designed lightbox which compensates for the lack of natural light outside. There is no need to look at the light; the user can carry out normal activities while in front of the box. Treatment is usually effective within three or four days.

Self-Inflicted Damage

I have already mentioned smoking in the context of air pollution. Let us have a closer look at it now.

When you actively smoke yourself or inhale tobacco smoke as a passive smoker, you expose yourself to over 200 chemicals which are produced with every single cigarette you light.

What happens when you inhale smoke is that your blood vessels constrict, and not just the ones in your chest, but every single blood vessel in your entire body! This leads to a deterioration of your circulation, often leading to cold hands and feet and, at worst, to a total collapse of blood supply in the extremities, leading to the necessity of amputating the limb concerned. But even if it does not come to that, the tightening of the blood vessels deprives your inner organs of oxygen and forces the heart to pump a lot harder to get the blood circulating through the system, and this makes you feel tired.

Another social drug that affects our energy levels adversely is caffeine, as found in tea, coffee, cola beverages, chocolate, cocoa and some pain-killers, but not in herbal teas. Initially it will appear to you that caffeine gives you a boost and heightens your energy levels because it produces a directly stimulating effect on the body through increasing the level of norepinephrine in the brain, making you feel more alert. Caffeine also activates your sympathetic nervous system by stimulating adrenalin release, but this boost does not last. With coffee you will experience a quick high followed by a rapid drop in energy, whereas with tea the boost-curve is slightly gentler, with the up less immediate and the low less sudden.

Nevertheless, as the artificial outpour of adrenalin into your system puts the body into overdrive mode, the natural balance is disturbed and a lot of the body's energy is squandered in dealing with the caffeine and all its side-effects, rather than doing the more useful job of keeping the immune system in peak condition.

Caffeine also upsets the blood sugar level as it forces the pancreas to produce insulin. Another disadvantage is that caffeine is acidic. This means that the body has to retain water to neutralize the acid, and this water retention will add extra weight to your body.

In order to re-establish a better physical balance, cut out caffeine from your diet as far as possible. You can gradually replace your daily cups of coffee or tea with decaffeinated varieties, or better still change over to herbal teas, dandelion, barley or malt coffee substitutes. Remember that every time you reduce your caffeine intake you are helping your body to recover its own natural strength which, if in good working order, is a lot more powerful and beneficial than cups of coffee can ever be.

ILLNESSES AND ALLERGIES

When you feel tired a lot of the time, this can be a side-effect of your body having to fight an illness or an allergy. It is always advisable to consult your doctor if you find that your energy levels have been low for longer than three months, just to reassure yourself that there is not a more serious illness such as anaemia or diabetes at the bottom of your fatigue. A thorough check-up will reveal whether your organs are working properly and whether there are signs of infection or fungal disease in your body.

Causes for tiredness can be manifold. Let us take a look at a few quite common conditions which are associated with fatigue.

Candida Albicans

Candida is a fungus which we all carry in and on us. It occurs in the lower bowels and on the skin and does not normally cause us any problems, provided our immune system is working properly. Candida is kept in check by 'friendly' bacteria in our gut so that it cannot grow out of proportion.

Problems arise when these 'friendly' bacteria in our colon are destroyed. This can happen when you are taking broad-spectrum antibiotics for a long time, for example if you suffer from acne or recurring cystitis. The antibiotics wipe out *all* the germs in your body, including the 'friendly' bacteria in your bowels! What these antibiotics do not do is wipe out the fungus, so that the Candida yeast can now grow uninhibited and cause symptoms such as oral and vaginal thrush, skin rashes, cystitis, fungal infection of the nails and infection of the tongue and throat, as well as indigestion, bloating and gas, and of course tiredness.

Another factor that can lead to Candida overgrowth is the use of steroids and the contraceptive pill, because these alter the hormone levels in the body, an imbalance in which Candida can thrive. In order to combat Candida successfully, it is essential to cut out sugar from your diet while also taking anti-fungal medication. (Although I do not usually advocate medication, it has to be said that in cases where your energy levels are very low and your body is unable to fight invading organisms by itself, medication has a useful temporary function.)

Avoid all sugar and sweetened foods, honey or foods containing syrup such as tinned fruit. Eat fresh fruit only in the morning and only on an empty stomach, with at least half an hour's break before eating any non-fruit food. Never eat fruit with or immediately after any other food.

If you are eating fruit on its own, at the right time of day, you are helping your body detoxify itself. As fruit passes through the stomach in 20 to 30 minutes (as opposed to other foods which take up to four hours to leave the stomach), very little energy is required for the fruit to be digested. Fruit goes straight into the intestines where it can release vitamins, minerals, fatty acids and amino acids into your system. Because the body needs to spend so little energy on digesting fruit, it can now get on with the detoxification process more efficiently.

It is also advisable to cut out milk and any milk products. They are difficult to digest and you absorb nutrients less well when your diet includes milk products, which means you get tired, even without suffering from Candida. Remember the glass of hot milk you were given as a child to help you sleep? It worked, didn't it? This is because milk releases a substance called serotonin in the brain which makes you sleepy.

Do not worry about not getting enough calcium if you cut out milk. Sprinkle a spoonful of sesame seeds over your salad every day and your body will be well provided with calcium.

Hypoglycaemia

Hypoglycaemia is quite common nowadays and has a lot to do with a low-fibre and high-sugar diet, but can also be exacerbated by stress.

Our bodies have two regulating mechanisms, an upper and a lower one, to maintain an optimum blood sugar level. We maintain a normal blood sugar level by eating energy-giving foods such as starches – as contained in bread, potatoes and rice – and sugar, as contained in fruits and honey. When a large amount of these carbohydrates are eaten, the upper regulating mechanism is set in motion, which results in insulin being released. If there is a long interval without food, the blood sugar level drops down to the lower regulating mechanism which will now set in motion a chain reaction of mobilizing the adrenal glands and getting the liver to release its stored sugar so that the blood sugar level can return to normal again.

When this mechanism is allowed to operate smoothly we will not have any problems, but if our sugar levels are artificially boosted by eating a sugar-rich diet, the blood sugar level shoots past the upper limit, producing too much insulin and then plummeting to below the lower level. This over-reaction can be brought about by consuming too much sugar, tea and coffee, by smoking, as well as by stress. One of the consequences of a steep drop in blood sugar level is tiredness because there is not enough energy-giving glucose in our system.

The easiest way to assess whether you suffer from hypoglycaemia is to check whether your energy levels improve as soon as you have eaten, and fall substantially 30 to 45 minutes after you have had your last meal.

MIND DEPRESSORS

Things that depress the body will always also negatively influence the mind. If your body is not working properly because, let us say, your liver and kidneys are functioning below par, you are unlikely to feel as bright-eyed and bushy-tailed as when your physical system is working 100 per cent smoothly. And of course, the same thing is true when it comes to the mind influencing the body. A traumatic event or a major change in your life will take strength out of you – all your mental energies are focused on getting to grips with the situation, and this automatically drains the body as well.

When you are under stress a lot of the time, no matter whether this is in your private or your professional life, it will eventually become visible on a physical level. We all know the dark rings under the eyes and the lines that appear on the face – and these are only the visible signs!

Because there is this interaction between body and mind, we must not underestimate how fundamentally our bodies are influenced by our minds. Our thoughts, reactions and attitudes towards life and people around us will be reflected in our bodies. We have a tendency to rationalize difficult events and situations to help us get to grips with them, but we might forget that our emotions can react quite differently and in a much less sensible way to those difficult events. When someone insults us, we may smile and give them a witty or disarming answer, so on the outside we seem to have coped well with the situation; but what about the inside? It is not a problem if the insult occurs only that one time, but what if you are working every day with someone who is rude and domineering? These psychological pressures can run your energy down just as a physical illness can.

PRESSURES FROM THE PAST

Often present-day stresses are felt more acutely because they occur on top of negative experiences from the past which have shaped our self-image and behaviour pattern in such a way that we become sensitive and anxious and therefore more prone to be defeated by current challenges.

These emotional legacies can hark all the way back into adolescence and even childhood. In my practice as an analytical hypnotherapist and psychotherapist I see people every day who suffer from fatigue and depression. In order to help them mobilize their energy levels again, it is often necessary to check back in time to see which past emotional events are blocking the free flow of energy.

... SO WHY NOT GET STARTED?

This, dear Reader, is probably the most personal chapter of the whole book. It deals with human resistance to do what you know would do you good. But now it comes to the crunch; we have to get started . . . Here are the excuses:

'All these suggestions sound very good, but I do not really have time to work on increasing my energy.'

This is many people's number one excuse. Is it yours? Think about it and be honest with yourself. How many things are you doing during the week that really have to be done? How many things are you doing that are unnecessary or not really important, such as watching TV?

And how many things could you leave undone or postpone until later because it will not really matter?

Remember: if you will not make time to stay healthy, you will have to take time out when you are ill.

'This is just not a good time for me to get started.'

This is another classic one. In principle you think it is all a great idea, but you have this one thing happening in your life which is so important that you cannot be distracted from it. Maybe you are working for an exam, or maybe the children are still too small, or you are in a very busy phase at work.

'I am doing so many things wrong, I am sure I would have to change my life completely in order to get any results, and I just cannot face that.'

Here is some good news – you do not have to change everything to get results. If you can just work on one thing, for example getting involved in an outdoor activity to counterbalance all the sitting around indoors at work, you will notice that your energy levels are perking up, both physically and mentally. It is not so important that you should correct all of your unhelpful habits. Just working on one is a good start. Do what you can do. This is a hundred times better than not doing anything at all.

'I feel a bit silly starting something new at my age.'

It may very well feel strange to do something new such as painting or singing, for example, especially if you have never really tried it before. In fact, the older you get, the stranger it feels, because the school days when you were learning new things all the time lie further and further back in time. Well, you are in for a very pleasant surprise, namely the experience of getting really wrapped up in a new venture, mastering a new skill. There is no need to worry about your age. If you are working on your energy alone at home, you can usually arrange it that you have some privacy, and if you are outdoors or joining others on a course, you will have the opportunity to observe that the younger participants need time to learn the new skill just as much as you do. Go for it! Once you have started, you will wonder why you did not do it before.

'Oh dear, I have broken my better-eating rules today! I might as well give up – I will never see it through anyway.'

This is not so much an excuse as an erroneous belief. When you have very high expectations of yourself and are also self-critical, it becomes hard to forgive yourself for not being perfect. And rather than being faced with your inability to do a new thing 100 per cent, you abandon your new venture. This is a great shame, because even improving on an old habit by 20 per cent is better than 0 per cent, isn't it? This book is not about being perfect. This book is about getting in touch with yourself and making a start into greater vitality and enjoyment.

'I hate feeling so exhausted after doing these aerobic exercises. I do not think this exercising business is for me.'

If you feel exhausted after exercising, you are doing something wrong. You should feel pleasantly tired, but never exhausted. There is a danger of pushing yourself too hard too fast in order to get results. Energy needs to be built up gradually, and each of us starts at his or her own individual level. Therefore there is no point in casting envious glances at other people who can do more in a shorter time. There is no way you can compare yourself with anyone else, because you are unique. The less you can make your energizing programme into a competition, the more fun it is going to be for you.

'I like the concept of eating healthily/developing a new interest/learning to relax, but I do not like the particular method that is described in the book.'

Some people want to eat more healthily, but they do not really want to stop eating biscuits. Some people like the idea of taking up a new hobby but they cannot warm to any of the examples described in this book. Some people are interested in the idea of relaxation, but they do not like to do it on their own. Fine, no problem! The suggestions in this book are meant to give you a guideline; they are not gospel. If you have different ideas of how to tackle the subject of relaxation, implement them. If you want to tailor my suggestions to suit your own needs, do so. Even if you still have the biscuits but eat a lot more fruit and vegetables, you are on a winning course. Some positive change is always better than no positive change, and as long as you get started, no matter in how small a way, you are already helping your body to raise its energy levels.

PART II

Physical Balance and Movement

In the days when people led more natural lives, their joints were well lubricated, their muscles strong and their heart and lungs resilient. Today we sit in cars and trains to go to work, we are hunched over a desk, often leaning forward continuously, and in the evening we slump in front of the television. This incorrect use of the body makes the muscles go weak and distorted, the blood circulation slows down and we feel continuously tired.

A heavy workload combined with insufficient time spent for eating, and living on processed rather than raw foods, aggravate the situation further. We lose touch with our bodies because we are too busy rushing around doing our job, and when we get a warning signal from our bodies we pop a pill for insomnia or take a powder for indigestion.

Our bodies work perfectly well as long as they are allowed to function as they should. We have an abundance of natural energy in us, provided we do not block that flow of that energy by overloading our bodies with the wrong food, with too much food or with distorting posture and lack of movement. If we want energy from our bodies, we need to give them back their balance.

FOOD

The quality of the food we feed into our system will determine to a considerable extent how healthy we are. Of course, there are individual differences between people. Some can tolerate a higher level of unhealthy foods than others; some put on weight more easily while others remain slim while eating larger amounts of food. We often measure health by the state of our body weight – fat is considered unhealthy, slim is deemed to be synonymous with fitness and energy. This is not at all so. While it is true that being overweight is an extra burden on your heart and your circulation, being slim does not mean that you are healthy.

Health depends on the quality of the food you eat. You can eat very little, maybe just a coffee and a bowl of cereal in the morning, two snack bars at lunch and a ready-made meal in the evening, and you may be very slim, but you are certainly not doing your body any favours. As long as you do not give your body some fresh foods such as fruit and vegetables, together with other foods high in nutritional value, you are depriving your body of the vital ingredients to keep your immune system working properly. We are today too caught up in the issue of weight to consider the far more important topic of body maintenance and prevention of illness. This is not to say that we should not watch our weight, but the reasoning for being slim needs to be understood as being a health issue rather than an aesthetic one. If you suffer from a lack of energy, the food you eat every day can have a lot to do with it.

EATING HABITS AND 'CLEAN' FOODS

When we are under any kind of stress, be it emotional or physical, we use up energy. Every single action we perform, be it lifting a heavy item or speaking, thinking or even sleeping, requires energy, but the activity that makes the greatest demands on our energy levels is the digestion of food. Even though we draw energy from foods, we lose some of it as the food is being processed by the digestive tract. The way in which we can reduce this loss to a minimum is by choosing foods that are as natural as possible and therefore clear through the digestive tract as effortlessly as possible. As these natural foods leave no wastes behind that the body cannot clear out easily, I call these foods 'clean'.

Clean Foods:
soya milk
fresh vegetables (raw)
fresh fruit (raw)

sun-dried fruit
brown/wild rice
raw nuts (almonds, cashews, pistachios, walnuts)
seeds (sesame, sunflower, caraway, poppy)
tofu (soya bean curd)
freshly pressed fruit juices
extra virgin cold-pressed oils
distilled water
coffee substitute (Caro, dandelion coffee)
herbal teas

OK Foods:
stoneground, wholemeal brown bread
rice cakes
unsalted butter
goat's milk
lightly cooked vegetables
poultry
deep-sea fish

juices made from concentrates
bottled and filtered water
honey

Energy-depleting Foods:
white bread
margarine
cow's milk and its products
ready-made dishes
cooked/stewed fruit
microwaved food
white rice
red meat (including sausages, bacon, salami,
    burgers)
juice 'drinks', canned and cola drinks
refined oils
tap water
coffee (including decaffeinated coffee)
tea
jams and marmalades

sugar
pastries and cakes
sweets
packaged breakfast cereals

If you make a list of everything you consume in one day,
including in-between snacks and any drinks, at least 80
per cent of your intake should come from the 'clean' and
the 'OK' sections. The more you can stay away from
energy-depleting foods, the more energy you are providing
for your body. Here are the reasons why so many foods
end up in the energy-depleting category.

Refined Sugar versus Fruit Sugar
Refined white sugar as found in chocolate, desserts, sweets
and soft drinks has been depleted of all its fibre, vitamins
and minerals through the processing procedure. You are
left with empty calories and excessive carbohydrates which
make you fat and produce acids in the body through
fermentation.

Sugar found in fresh fruit, however, provides you with natural, untampered fructose which can be turned into energy-giving glucose easily and passes into the bloodstream quickly.

Cow's Milk and its Products

The enzymes necessary to digest milk are renin and lactase. In most people, these enzymes are more or less gone by the age of three. Moreover, milk has casein in it which coagulates in the stomach and adheres to the lining of the stomach and intestines, preventing absorption of nutrients into the body. This mucus-forming property of milk can also show itself in symptoms of a clogged-up nose and ears.

Margarine versus Butter

Contrary to advertisements, margarine clogs up the arteries. The process of making margarine involves turning polyunsaturated oil into fat by blowing hydrogen gas through the oil until it solidifies. The process of hydrogenation actually converts the originally polyunsaturated fat into saturated fat, which causes a rise in cholesterol. Any artificially added vitamins do not really make it any healthier.

Butter may be animal fat, but at least it is natural rather than synthetic, and it is neutral rather than acid-forming, unlike all other dairy products. The body needs fat, so the addition of modest amounts of butter are fine for a healthy diet.

Meats

Meat is concentrated protein food, and protein is the hardest food to digest when it comes from an animal source. It can take over two days for meat to pass through the entire gastro-intestinal tract, and this takes up a great deal more energy than any other food. A much better source of protein is tofu because it is vegetable-based and easy to digest.

Also, when we eat meat we are at the end of the food chain. As chemical fertilizers and pesticides are used on crops and fields, these chemicals find their way into the bodies of animals who feed on these crops. The chemical contaminants are consequently stored in the animals' body fat so that the pesticides become more concentrated. When we eat animal flesh, we not only eat the hormones that may have been fed to the animal, but also a heavy dose of fertilizer and pesticides!

Soft Drinks versus Freshly Pressed Juices
Soft drinks contain acids which remain acidic as they go into your body. In freshly squeezed citric juices, the acids will turn alkaline in your body. In the manufacture of soft drinks, however, the artificially added citric acid is usually extracted with heat from its original source and thereby fractured, and will therefore no longer convert to alkaline.

In addition, soft drinks contain a lot of sugar, and it does not really matter whether it is normal sugar or a sugar substitute; they are both equally harmful, so do not be lulled into a false sense of security by choosing the 'diet' variety!

Breakfast Cereals

In the production of cereals, lye is used (which is a caustic substance), and up to 50 per cent (often more) is pure sugar, with another 30 per cent or so of refined flour. The rest is saturated fat, salt and some inorganic vitamins and minerals. There is hardly anything of value in these cereals.

Processed Foods

When food is processed it undergoes a number of artificial processes such as heating, fragmenting, pulverizing or changing in any other unnatural manner. Then chemicals are added to enhance the appearance of the new product, to keep it soft, stabilized, firm or moist. Then artificial nutrients are added to make up for everything that was destroyed during processing.

The word 'natural' on a packet does not necessarily mean you are eating a food as it was designed by nature. 'Natural flavouring' can still mean that the flavour has been artificially produced, but in a way that chemically mimics nature.

The only really natural food is the one that comes from trees, bushes, and out of the earth. Processed food is not natural and should be avoided as far as possible. This also goes for processed vegetables.

Refined Foods
White bread, white rice and refined oils have all been stripped of their original goodness. Refining is really just another word for processing. As a rule of thumb, stick to those foods which are as close to their original state as possible.

DETOXIFICATION

When you feel sluggish or uncomfortably heavy, this is a sign that your body is carrying unnecessary toxic baggage. As long as this waste is allowed to stay in your system, your energies will be squandered – the body is busy trying to eliminate the waste, and this means there is no energy left for anything else. So if you want your energy back, help your body rid itself of wastes.

With detoxification or fasting, it is very important to take things gently. If you are physically or emotionally robust you can proceed faster than if you feel delicate.

Always speak to your doctor first to check whether your chosen detox programme or fast is appropriate for you. As a rule of thumb, it is better to detoxify gradually rather than in one big go, especially if your energy levels are very low to start with.

There are some points worth considering before embarking on your detox programme:

- If you are aiming for a one-day or two-day fast, be aware that it can take several months until your body has expelled all the toxins, so you will have to repeat the fast once a week for a while. Do not forget that you have accumulated these wastes over many years, so you cannot expect them to be flushed out in one weekend.

- Be prepared to get worse before you get better. You may experience headaches, a furry tongue, bloating and irregular bowel movements initially. These are slightly unpleasant but positive signs that your body is tidying up on the inside. The more toxins you have in your body, the more likely you are to experience any of these side-effects.

- Choose a weekend for your detox programme or, better still, a holiday period. Avoid all social engagements, especially the first few times, as you cannot be sure how you will feel.

- Put the answering machine on, get out some soothing music, make sure you have some relaxing reading material. Some people find they sleep a lot during their fast; that is fine too.

- Alternatively, you can go for a gentle walk or sit out in the garden.

- Keep your fluid intake high. This helps the liver and kidneys work better and detoxify more easily.

- Allow one day of re-adaptation to normal eating for every day you have been fasting.
- Keep warm.

Water Fast

Drink at least 4 and at most 8 pints (2–4 litres) of water during 24 hours. Squeeze some lemon juice into the water to give it a slight flavour if you want to. Use still rather than sparkling water. Alternatively, drink the same amount of herbal tea, cooled down. Good varieties for fasting are linden blossom, nettle, chamomile, sage and lemon verbena.

Fruit Fast

You can have up to 3 lb/6.6 kg of either grapes, apples, pears or papaya. Wash the fruit well and eat it with the skin (except papaya) to give your body roughage. Eat slowly and chew properly before you swallow. Eat your chosen fruit every two hours and drink a maximum of

4 pints (2 litres) of water, but never with the fruit. Drink
15 minutes before eating and leave a gap of 30 minutes
before you drink after eating.

Salad Fast

Eat only salads, raw vegetables and sprouting seeds. Celery
is particularly valuable as it has high cleansing properties.
You can add a little lemon juice or virgin olive oil for
taste, but no salt or pepper. Drink herbal tea, cooled
down, or water, but not more than 2 pints (1 litre).

Start with just one day of detoxification and repeat the
fast once a week for at least two months. If you do your
fast on a Saturday, you will have the Sunday to re-adapt to
normal eating. It is essential that you do not go back to
normal eating on the Sunday but gently ease your body
back to a greater amount of food.

Re-adaptation Day

Eat fruit in the morning and raw vegetables in the
afternoon. In the evening, prepare a light vegetable meal,

stir fried or lightly boiled. You can add some extra protein by including tofu. Drink 2 pints/1 litre of fluid, either water or herbal tea cooled down.

ALLERGIES

Food allergies can make you feel depressed and tired, give you migraines and nightmares, cause fits, convulsions and fainting, give you a swollen face or limbs, and cause rashes, eczema and many other symptoms, both physical and psychological.

When I speak of 'food' allergies, this includes allergies to food additives such as preservatives, colourings and flavour enhancers, as well as drinks, including tap water. You can have an allergy to just one foodstuff or to several. Even though you may think of some foods as harmless, you should not exclude any foods from your search for the allergy-causing agent. I know people who are allergic to celery and cucumber!

If neither wheat nor milk causes you problems, keep looking for other culprits. The best way of establishing which foods give you an allergic reaction is to start a food diary. Write down *everything* you eat and drink every day, including snacks and sweets, for four weeks. Also note down how you feel after having eaten these foods; notice any connections. Do you feel unwell every time you eat this food, or can you sometimes eat it without feeling ill? How long after eating do the ill-effects occur?

You do not have to keep an elaborate diary; a piece of paper will do. Carry it around with you and jot down what you eat and drink right after having consumed it, as you may have forgotten some of the items by the evening, especially if you have had a busy day. Use abbreviations for various foods and symptoms, but make sure you use them consistently. Hang on to these notes; do not throw them away. You may be going through a good patch at the moment, but you will want to be able to compare what foods you were eating when you felt good to those you are eating when you feel unwell. Keep wrappers and labels;

this will relieve you of having to write down all the ingredients of a ready-made dish.

If you suspect that you might have an allergy and would like expert help with it, you will find that there are a number of specialists you can approach. You can either consult a medical doctor who specializes in allergies, or you can go and see a dietitian, nutritionist, homoeopath, herbalist or naturopath, all of whom will ask you a great number of questions which will help them assess which foods you should avoid. Alternatively you could see a health kinesiologist, who will establish what you are allergic to by muscle testing, a simple but highly accurate diagnostic tool (provided you consult an experienced practitioner).

VITAMINS AND MINERALS

Vitamin and mineral supplements are nowadays widely available in shops and can be bought without prescription.

But are they really necessary, or is it all a big marketing hype to entice consumers to part with their money? Some schools of nutrition claim that as long as you avoid meat and eat raw fruit and vegetables, you will get all the vitamins and minerals you need. However, this theory does not take into account that we are today exposed to a lot more pollution than we ever were, and this increases the body's demands for vitamins and minerals to help fight off the invading toxins.

If your body is to stay healthy, it needs to produce millions of new cells every week. In order to do this regenerative work and keep the immune system functioning efficiently, you need a diet which is rich in vital nutrients – and raw fruit and vegetables are certainly the best source. However, some circumstances can mean that you are depleted of certain nutrients. If you smoke or drink, if you are under stress a lot of the time, or if you are presently or have been in the past on antibiotics, corticosteroids, diuretics, anti-inflammatory drugs or on the contraceptive pill, even a healthy diet cannot supply your increased needs, and you will have to supplement your diet with extra vitamins and minerals. Many women are deficient of magnesium, which is lost every month during menstruation. This mineral is needed by the body to convert blood sugar into energy, and it helps with the uptake of calcium into the bones, besides having a calming effect on the nerves. Even on the best diet you will only obtain 300 mg, whereas some women need 400–600 mg daily.

Here are the minimum daily amounts of vitamins and minerals that you should get – please note that the amounts indicated refer to someone in good health with a normal energy level. If your energy levels are low, you will have to exceed these recommended amounts, as explained in the following section.

When suffering from lack of energy, supplement the following vitamins and minerals:

- Vitamin A (retinol)

  Needed for maintaining mucous membranes, skin and cell membranes; helps protect against infection.

  Natural sources: fish-liver oils, carrots, sweet potatoes, spinach, pumpkin, broccoli, watercress, asparagus; watermelon, apricots, peaches, nectarines; eggs.

  To build up energy take 7–15,000 iu per day. *Do not exceed 20,000 iu.*

- Vitamin B₅ (pantothenic acid)

  Needed for proper functioning of adrenal glands and for producing antibodies. Helps with allergies and arthritis.

  Natural sources: broccoli, cabbage, cauliflower, peas; sesame seeds; whole grains; eggs.

  To build up energy take 300–1,000 mg per day.

- Vitamin B_{12} (cobalamin)

  Needed for red cell production and for the smooth functioning of the nervous system. Deficiency can lead to fatigue and confusion.

  Natural sources: offal, fatty fish, eggs.

  To build up energy take a good vitamin B-complex which is yeast free and contains all the B vitamins (B_1, B_2, B_3, B_5, B_6, B_{12}, folic acid, Para Amino Benzoic Acid [PABA] and biotin).

- Vitamin C

  Protects against effects of immune-suppressing hormone (corticosteroids) which are produced under stress. Vital for continuous repair and regeneration of body tissues.

  Natural sources: green peppers, broccoli, spinach, tomatoes, parsley, kohlrabi; citrus fruit, strawberries, kiwis.

To build up energy take 3,000–5,000 mg per day. Start with 500 mg in the morning and 500mg in the evening and build up until you reach your limit (when you experience runny stools). Cut back to 1,000 mg less than that last dosage. *Do not exceed 9,000 mg.*

- Vitamin E (tocopherol)

  Contributes to healthy muscles and nerves, essential for proper wound-healing.

  Natural sources: soya beans, cold-pressed oils, broccoli, Brussels sprouts, spinach; pecans, walnuts; whole grains; eggs.

  To build up energy take 100–200 mg.

- Magnesium

  Vital for normal cell function, transmission of nerve impulses and for muscle contraction and relaxation.

Natural sources: spinach, Brussels sprouts, broccoli, corn; figs, lemons, grapefruit, apples, almonds, sesame and sunflower seeds; whole grains; honey.

To build up energy take 500 mg per day, together with the same amount of calcium to prevent depletion. Reduce to 300 mg once there is an improvement.

- Iron

  Acts as oxygen-carrier in red blood cells and oxygen-reservoir in muscles, as well as helping to resist infection.

  Natural sources: spinach, beans, peas, Brussels sprouts, watercress, artichokes, broccoli; watermelon, dried fruits (prunes, apricots, dates, raisins), nuts; sesame and sunflower seeds.

  To build up energy take up 24 mg per day in its ferrous form so it is compatible with vitamin E.

- Selenium

  Assists the excretion of heavy metals from the body.

  Natural sources: onions, garlic, tomatoes, broccoli, asparagus, cold-pressed oils; whole grains; most nuts; tuna fish.

  To build up energy take 200 mcg. *Do not exceed this dose.*

It would also be helpful if you could have your doctor or nutritionist check whether you lack digestive enzymes or hydrochloric acid. Both can cause poor absorption of food, which means you are not getting all the nutrients out of the food you are eating. Both conditions can be easily amended by taking supplements of digestive enzymes/hydrochloric acid (sold in tablet form) before or during a meal.

TAKING LIQUIDS

Besides air and food, water is one of the necessities of life and is of vital importance to our health. Without water, our energy levels drastically diminish. In order to maintain good health we need to drink six to eight tall glasses of water a day, which is between 1 and 2 litres (2–4 pints).

One of the places where the body gets water from if you do not drink enough is the colon. This dries up the colon area and brings on constipation. Digestion and bowel function are dependent on the lubricant water.

Drinking six to eight 8-oz glasses of water is sufficient for the average person. If you are overweight, you should increase your water intake to nine to ten glasses. Also, when you are exercising or when the weather is hot and dry, increase your water intake.

As you are drinking more water, make sure the quality of your drinking water is good. Avoid tap water if you can. It contains chlorine, which is a bleach, and it often also carries other chemicals which have escaped the cleaning process. Drink bottled water instead, or have a carbon filter installed under the sink. Use filtered water for cooking and washing vegetables, as well as for drinking. Cold water appears to be better than warm or hot water for increasing energy levels; it can also help burn calories better. But stay away from ice cubes when you go out – they are normally made with tap water. Still water is preferable to carbonated, as the gas bubbles can upset your stomach if you are sensitive.

Do not drink during meals because the water will dilute the gastric juices and inhibit good digestion. Drink either 15 minutes before a meal or leave a gap of two hours until you take liquid after a meal.

Tea, coffee, alcohol, lemonade, fruit and vegetable juices are no substitutes for water! Tea, coffee and alcohol dehydrate the body, as they act as diuretics, so avoid these – or if you cannot live without them make sure you still have *at least* eight glasses of water to make up for the dehydration.

POSTURE

When I went to my Alexander Technique teacher Carmen for the first time because of my persistent lower back problems, one of the things she asked me to do was to stand straight. I lifted my chin up, squared my shoulders and did my best to stand tall, sure of my knowledge about what was 'straight'. Then Carmen started to rearrange practically everything about me – she lowered my chin again, gently pressed on my hip bones to tilt my pelvis forward, and she also made me unlock my knees so that they were coming a little bit forward. After all this, she carefully readjusted the tilt of my body by guiding me to lean further to the left, which made me feel a bit confused because I felt as if I were in the process of falling over to the left any moment. 'And this,' Carmen said, 'is straight!'

Take the time to observe your own posture. Notice how you sit when you are answering the phone, driving your car or reading or relaxing. Is your back rounded? Do you twist your legs around one another? Are you bent over a

desk all day, cocking your head to one side trying
to hold the phone against your shoulder? If you do,
you may already have noticed some warning signs –
backaches, headaches, constipation, poor circulation
and fatigue.

BODY KNOW-HOW

Good upright posture allows every part of the body to
work well. Here is an easy way of checking that you are
standing with your body properly poised:

- Stand with your feet just slightly apart.
- Sway very gently forwards and you will feel the
  pressure of your weight on your toes. This is too
  far forward. Now sway very gently backwards so
  that the weight is on your heels. This is too far
  back. Now sway very slightly forward again so that
  the weight lies just in front of your ankles. This is

where it should be. (Do not forget to breathe while you are doing this!)

- Now move your attention to your knees. Make sure your knees are not locked back. If you can wiggle your knee-caps up and down, they are relaxed. (Make sure you breathe while you are making these adjustments.)

- Now move up to your pelvis. Gently move your pelvis forward, with your hands on your hips. You can feel that this puts your body out of alignment. Now push the pelvis backwards so that your bottom sticks out – obviously wrong. Now readjust your pelvis to the middle, imagining your body weight passing through your knees down to the front of your ankles. (Breathe while you are doing this.)

- Now stretch your spine and neck gently upwards, eyes facing forwards, chin at right angles to your neck. Allow your arms to hang loosely by your sides.

You have now not only eased the strain on your ligaments and bones, but also helped your chest to expand so that breathing becomes easier and the liver has more space in your body to do its work efficiently.

Try these adjustments in front of a mirror, standing sideways to it; you will be surprised how your shape alters with every correction you make. There is no need to maintain the corrected posture for very long. Practise it often though, such as when you are standing in a queue, or during a break at work. It will soon become automatic.

GIVING YOUR BACK A REST

To keep your back in working order, it needs to be allowed to rest. While you are sitting or standing up, gravity puts pressure on the discs and squashes them. Muscles become shorter, joints are jammed together and fluid is squeezed out from between the intervertebral discs. If you were to measure your height at the end of a busy day, you would find that you had shrunk by up to 1 inch!

The function of discs is to give resilience and elasticity to the backbone, but in order to be strong the discs need to have a 'cushion' of fluid in their centre. Allowing the spine to lengthen by lying down during the day will allow the fluid to be reabsorbed into the discs. It also has the added benefit of allowing the back muscles to loosen up and smooth out, which helps the back recharge energy. Here is an exercise that will rest your back correctly:

- Lie down on your back, upper arms resting comfortably on the floor and your hands loosely resting on your lower abdomen.

- Your head should be supported by a towel that is folded to bring your head far enough off the floor so that your head is neither tilting backwards nor so high that your chin is pushed towards your chest.

- Now pull up your legs so that your knees point to the ceiling. Keep the knees apart (about shoulder-width apart) and keep your feet flat on the floor.

You will feel how this brings your lower back closer to the floor.

- Rest like this for about ten minutes every day. It will not only help your back to relax, but also your mind.

EXERCISING

You may be surprised at the type of exercises which are recommended in this book. Walking, stretching and breathing may appear to you to be truly 'lame ducks' if the aim is to produce energy.

The reason behind my selection is that these are fundamental exercises which can be done by virtually anyone, no matter what level of fitness. These exercises are also more powerful than you may think and the good thing about them is that they can be practised by any age group. Even if you have never done a stroke of exercise since you left school, even if you are 85 next week and you have arthritis, you can still start today on these particular exercises. If, on the other hand, you are reasonably fit and want to give your energy levels an extra boost, you can make these exercises much more demanding. It is not necessary to push yourself hard to get results, but if you are keen to push there is a lot of scope for doing so.

There are many myths going around in connection with lack of energy and exercise. ME sufferers are told to rest so that they do not make their condition worse. Others with less serious energy problems might have tried an exercise class, only to feel absolutely shattered at the end of it, unhappy that they could not keep up with the others in the class and consequently feeling like a failure. It is true that if you suffer from ME, also known as Chronic Fatigue Syndrome (CFS), any bursts of energetic activity will result in even greater fatigue which requires you to rest for several days to recover, and this is why most doctors advise sufferers to rest a lot. The problem with prolonged rest is that it makes your body weaker;

muscles waste away and your active lung capacity is reduced, making you more vulnerable to further illness. Resting is a must during and after any illness or viral infection, until the illness has abated. However, prolonged rest is a hindrance to full recovery because it reduces your tolerance for activity. The longer you stay inactive after an illness, the greater the fatigue becomes.

When this period of prolonged rest is followed by a sudden burst of activity, this will naturally result in exhaustion, simply because the body is overtaxed compared to its previous level of inactivity. This stop–start pattern means that sufferers are unable to build up a sustained level of recovery, and this can be very demoralizing.

This unsatisfactory outcome of any physical activity makes many people think that exercising actually exacerbates their condition and that 'too much' might cause permanent damage. Even doctors will support this belief. Consequently, the sufferer becomes more and more restricted in his or her 'action radius'.

It is often people who tend to overwork and push themselves very hard who are prone to exhaust their energy resource to a point where their body just gives up. These personalities are particularly annoyed by their enforced inactivity and will try too early to become active again. They push themselves into excessive activity and exercise, and the ensuing bout of deep exhaustion makes them feel even more exasperated with themselves, so they push some more … From my experience at my practice, I have found that most ME sufferers are hardworking, ambitious people who find it difficult to be patient with themselves. This is often what made them ill in the first place. They will then quite naturally overwork again when they set about tackling their energy problem.

Exercise is perfectly all right, even with ME, *as long as it is started on an extremely low level.* Activity needs to be re-introduced gently to the body. Muscles need to be built up slowly, and active lung capacity needs to be developed at a very gentle pace. New exercise levels should be maintained for at least one week until they are increased.

As an example, if you have not exercised at all because of a recent illness, start with as little as half a minute's stretching or walking three times a day, building up to a minute over the next two weeks. This may sound a negligible amount of physical activity, but it is the only way in which you can strengthen your body cells steadily and maintain the improvement.

Even if you are not an ME sufferer and only very low on energy, you should observe this rule of a slow build-up. Forget everything you have ever heard about resting pulse and aerobic rate – whatever physical activity you do regularly, no matter how gently, will have a beneficial effect on your energy levels.

WALKING

'I have two doctors – my right leg and my left leg.' This old Yorkshire saying has a lot of truth in it. Walking is an excellent medicine for building cardiovascular and respiratory fitness and protecting you against heart disease, cancer and other illnesses. Researchers at Loughborough University have discovered that walking not only reduces blood-pressure, but also reduces the risk of clogged arteries. Volunteers who took an energetic two-hour walk the day before a high-fat meal had 30 per cent less dietary fats in their blood than when they ate a less fatty meal but rested the day before.

Ideally you should walk briskly, which means covering 3.5 to 4 miles an hour. If you were to walk for one hour at this speed you would burn up approximately 400 calories, more if you had some uphill inclines as part of your route. However, this sort of tempo is not recommended if you are recovering from an illness or if your energy levels are very low for any other reason. Remember that anything other than lying down or sitting around will challenge your body.

Walking is an excellent antidote to depression and insomnia. It increases the metabolic rate and helps you keep weight off; it also increases your intake of oxygen, which supports your immune system because bacteria can only thrive in tissues which are low on oxygen.

STRETCHING

Stretching is seen by many as something you briefly do before or after an exercise routine, before you finally relax. It is mostly treated as an addendum to running, jumping or any other athletics or aerobics or as part of the cool-down after exercising. As such, stretching is not considered exercise at all. Perhaps this is because, when you stretch, you do not move quickly and you do not seem to be 'doing' anything spectacular. Many people also consider stretching to be less important because it 'only' gives you more flexibility, which is not commonly associated with the creation of energy.

The truth is that stretching is a vastly underrated form of exercise which deserves to be recognized as an independent discipline with enormous benefits. It not only creates a great improvement in energy and physical well-being, but also diminishes stress and lifts your mood. You will find it easy to concentrate, your breathing will become deeper and you will feel relaxed after a good stretching session, without being tired. Hatha yoga is based on a great array of stretching positions, and anyone who practises this type of yoga regularly will be able to confirm how immensely it contributes to physical and emotional well-being.

During stretching, the body is being brought back into alignment through balancing the muscles, correcting the tilt of the pelvis and easing the spine's curves. Stretching allows the chest area to expand, so that organs operate better as blood and lymph circulate much more freely and are able to access areas of the body which beforehand were cramped up with tight muscles. This also stimulates digestion.

When you have been inactive or overworked for a while, your body shrinks. Not only do your muscles lose their tone and become shorter, but your joints are also jammed together and fluid is squeezed from between the intervertebral discs. Basic daily stretching strengthens the spine, which is vital for the full use of our limbs. Think of the spine as a central support around which all other movements happen. The more supple this rod is, the greater the elasticity of the rib muscles and the diaphragm. The spine is not a solid piece of bone but a column of 33 sections called vertebrae. These vertebrae naturally form a curved column rather than a straight one. The natural curvature of the spine gives it greater strength, far greater than if the spine were rigid. This is where stretching becomes important: it keeps the spine and the muscles around it flexible and strong. This in turn exercises the inner organs indirectly as increased blood flow and more oxygen can reach them, all of which has a massaging effect.

Here are some simple stretching exercises which you can do, no matter what your energy levels are.

Exercise 1: Standing Stretch

- Stand with your feet together, toes pointing forward. If you are not very steady on your feet, have them slightly apart. Let your arms hang by your sides, fingers pointing down. Keep your chin at a right angle to your neck, eyes looking forward.

- Take a deep breath in and, as you are breathing out, stretch as follows: Keeping your heels and toes firmly on the ground, try to feel as if you were extending your legs towards the ceiling, lengthening your spine upwards as if you were trying to straighten out the curves of the spine all the way up through your head, lifting the base of the skull. At the same time, drop your shoulders while stretching your arms down into your fingertips.

- Do not forget to breathe!

Exercise 2: Stretching Up

- Stand with your feet together.
- Interlock your fingers behind your neck, elbows pointing outward.

- Breathe in, and as you are breathing out, turn your palms upwards and push your arms up, at the same time time tightening your legs and the trunk of your body as you did in the standing stretch.

- Do not forget to breathe!

Exercise 3: Sitting Stretch

- Sit on a blanket on the floor or on a firm bed, legs in front of you. Place your hands with your fingers pointing forward near your hips.

- Breathe in, and as you breathe out hitch up your feet, pressing your thighs, knees and calves down. At the the same time, press down onto your hands while you stretch the trunk of your body upwards, as in the standing stretch.

- Breathe while you are holding the stretch!

Exercise 4: Leg Stretch

- Stand with your feet about 3 ft (1 m) apart, toes pointing forward.

- Turn your upper body to the right and let your legs and feet follow, turning 90 degrees right so that you are in a 'walking off to the right' position.

- Bend the front leg at the knee and stretch the back leg while keeping it straight, heel on the floor. You should now feel a stretch through the back of your calf muscle.

- Return to the initial position, then turn to your left and repeat, stretching out the back of the other leg.

- Make sure you breathe while doing this!

BREATHING

Breathing is normally controlled by our autonomic nervous system. Just like the beating of our hearts, the functioning of our sweat glands and the contracting and dilating of our pupils according to the intensity of light they are exposed to, breathing is regulated without us having to attend to it consciously.

Unfortunately, most of us do not breathe deeply any more. Stress and emotional upheaval can set up an unsatisfactory breathing pattern where the breath is held a lot, then quickly released; then another shallow breath is drawn in, which is then held in turn until the body's survival mechanism will insist on another breath. If you have ever seen a child who is upset and crying, you will have noticed the irregularity of his or her breathing when the child tries to speak while upset – the words come out in 'gusts' together with the sobs.

Anxiety, fear and a lack of confidence are emotional states which go hand in hand with a faulty breathing pattern. Panic attacks are a classic example of this. They make you feel rooted to the ground, unable to move, your heart racing, cold sweat breaking out and the sensation that you are going to faint or even die. These physical symptoms occur as a result of an emotional stress overload which leads to rapid shallow breathing.

This type of breathing causes palpitations and a feeling of faintness as both the heart and the brain are deprived of oxygen. Panic attacks can be so serious that the sufferer may suspect that he or she has had a heart attack, when in reality it is only great physical tension and consequent lack of oxygen that have created the symptoms.

A good breathing pattern is regular and comes from the abdominal area. As you breathe in, the expanding lungs push down the diaphragm – a dome-shaped muscle which is situated just below the bottom of the lungs. As the diaphragm is pushed down, it causes the abdominal muscles to relax and rise, which in turn allows the lungs to expand even further. On the outbreath, the diaphragm relaxes and the abdominal muscles can now contract to push out air. This means that you will experience an extending of your belly when you breathe in and a deflation of your belly muscles when you breathe out.

Exercise 1: Abdominal Breathing

- Lie on your back, knees bent and both feet on the ground. Close your eyes.

- Place one hand on your belly area, just above the navel.

- Breathe through your nose as you would normally do and feel what is happening. Is the hand on your belly rising as you breathe in, or is there no movement there? If your belly area is rising on inhaling, your breathing pattern is correct. If your hand is not rising, work on the next steps.

- When you breathe in, make a 'fat' belly. Consciously push out your belly muscles. It is OK to do this on purpose, even though it is not natural. You are just doing it at the moment to re-educate yourself. On breathing out, pull in your belly muscles strongly.

- Let the belly muscles relax.
- On breathing in the next time, you will notice how your body adapts to the right breathing pattern automatically – belly coming out as you inhale, and deflating as you exhale.
- Continue to breathe in the correct way for a little while so you can become used to the new pattern.

If you have a tendency to tighten up and hold your breath a lot, use the following exercise to free the abdominal area from stress.

Exercise 2: Loosening Up

- Sit or stand up. Stretch your arms up. If you feel very weak, you can also do this exercise lying down, but still lengthen out both your arms, allowing them to rest behind you.

- Breathe in through your nose and make sure your belly muscles extend as you have learned in the previous exercise.

- Emit one sharp outbreath through the mouth on a 'HA!' sound, as if you wanted to breathe on a window before polishing it, but in a much sharper way. As you expel the air in this abrupt manner, your belly should sharply tuck in.

- Let your arms come down to your sides and take a few normal breaths, then repeat the exercise two more times.

PART III

Resting

RELAXATION

The boss is in a bad mood, the children are misbehaving, you have made a mistake which you will have to own up to, the train is late and overcrowded – all these everyday pressures take it out of you and make you feel jaded at the end of the day. If these types of pressure, together with other stresses, continue over a period of time they can seriously drain your energy unless you know how to recharge your batteries and counterbalance this drain.

Proper relaxation (both physical and mental) is a skill that can be easily learned, and it is worth investing a little time in exploring which type of relaxation exercise works best for you. If you are impatient with yourself, start with one of the shorter ones – the approximate time each exercise takes is given along with each one.

No matter which method of relaxation you choose, you will be able to feel some of the physical benefits quite distinctly. As you unwind, a number of body processes begin to slow down – your heart rate decreases, blood pressure falls and breathing calms down. At the same time, other processes are stimulated, such as the flow of saliva and the release of bile, which helps break down fats and improves digestion. Any physical processes can now function much more easily. As the body turns to its natural balance during and after relaxation, the mind and emotions follow suit. Apart from feeling calmer and more composed, you are also mentally more agile. Your concentration and memory improve, you feel stronger and more centred, your mood lifts and problems are put into perspective.

Although it is comparatively easy to learn to relax, it requires a certain amount of discipline and dedication to integrate it into your life. It is worthwhile releasing tension on a regular basis; as you are letting go of annoyance, frustration and fear, your body and mind can simmer down and recoup lost energy.

Find a quiet, comfortable place to relax. Ask your family not to disturb you unless the house is on fire. If you live on your own, put the answering machine on and turn the volume of your phone off (put a note by the phone to remind you to switch it back on again afterwards).

Having said this, I would like to start off with two exercises which do not require you to shut yourself away. They are very quick ones and useful for situations where you have no opportunity to do an 'official' relaxation exercise or where it would be inappropriate to do so because there are other people around.

Exercise 1: First-aid Relaxation (30 seconds)

- Take a deep breath in, hold it and tense as many muscles as possible. Clench your teeth, tense your shoulders, clench your fists or, if other people are around you, tighten up your fingers by lengthening them out. Tighten your legs and your feet.

- Hold the tension for a moment, then breathe out and release all the muscle tension.

- Allow a few normal breaths in between, then repeat steps 1 and 2 twice more.

Exercise 2: Quick Visual Relaxation (1 minute)

If anything is annoying or frustrating you, close your eyes and imagine an old-fashioned stone bridge which spans a pleasant river, with grassy banks on both sides. Imagine a piece of wood being carried along the river, and imagine that this wood represents the reason for your current annoyance. Now watch the piece of wood drift along the river, under the bridge and off into the far distance until you cannot see it any more, and think to yourself, 'Next week/month/year, this matter will all be water under the bridge.'

For the rest of these exercises, you might find it helpful to lie down on a firm bed or sofa or on the floor on a blanket with a cushion under your head. Make sure you are warm enough – it is very difficult to relax when your feet are turning to icicles!

Exercise 3: Progressive Muscle Relaxation (5 minutes)

- Close your eyes so you can concentrate better on what you are doing.

- Start off by slowly tensing the muscles in your feet until you have a clear feeling of tension in them. Hold the tension for a moment so you can be fully aware of it. Now let the tension go slowly, relaxing the toes, instep, heel and ankle. Be aware of the feeling of relaxation and stay with that awareness for another moment.

- Now concentrate on your calf muscles. Tense them up slowly and, while you do so, try not to disturb other muscles. Hold the tension for a moment. Now release the muscles again slowly and feel all the muscle fibres, joints and tissues relax.

- Continue in the same manner in the following order: thighs, belly area, chest area, hands, arms, shoulders and neck, and face (frown and grit your teeth).

- Once you have tensed and relaxed all the muscles from the tips of your toes to the top of your head, be still for another few minutes and enjoy the feeling of inner tranquillity and well-being.

Exercise 4: Daydream Relaxation (5–10 minutes)

- Imagine that you have won a great deal of money; so much that you do not have to work any more. (If your main stresses have to do with money and/or work, you will probably find this first sentence already immensely relaxing!)

- Make yourself comfortable, close your eyes and let your mind produce all the images that, for you, go with your new financial status – decking yourself out with new clothes, travelling, buying that sports car, buying that big house, having others pander to your every need, and so on.

- Go into great detail about every aspect of your venture. If you think about buying a new house, imagine exactly what the house looks like, and make the whole enterprise into a story – see yourself looking into the estate agent's window (looking only at the photographs of the houses and not at the prices, because they do not matter anyway); being shown around the property, moving in, having it decorated (someone else is supervising the work and yelling at the workmen if necessary), and then the house in its finished state.

- Spin it all out as long as you can; the longer you can stay with your daydream the deeper your body and mind can relax. This pleasurable escapism will leave you refreshed and possibly motivated to earn the money to fulfil some of your more expensive dreams . . .

SLEEP

If you have ever suffered through a few interrupted nights in a row, you know what it feels like to wake up tired with a head full of cotton wool and just wishing you could turn over and get back to sleep. You start your day with your energy levels down to zero, and even though you eventually get your act together and trot off to work, after lunch at the latest the tiredness overcomes you again. You bulldoze through it, only to find that you are wide awake again in the evening when it is time to go to bed.

Another way in which interrupted sleep can affect you is by sending you into further overdrive during the day so that you never feel tired as such but are always on edge and uptight. This can become a vicious circle – you are stressed in the first place and therefore unable to sleep soundly, and then the lack of sleep aggravates the stress even further. Even when the original stress has abated, we are often still stuck with insomnia which can then become a habit.

In most cases, people find that they can go back to a good night's sleep once the original worry about something at work or at home is over; other people are prone to tension because of their personality make-up or through the way they lead their lives.

To this day, there is still no definite answer as to how many hours of sleep is normal. Variations range from happy five-hour sleepers to people who swear that they need nine hours' sleep or they will not be ready for the day. The majority of people seem to feel refreshed after seven to eight hours, but that does not mean there is something wrong with you if you deviate from this average number of hours.

Here are a few things to consider if you would like to improve your sleep in a natural manner:

- Keep your bedroom cool, but make sure you are comfortably warm in bed.
- Check whether your mattress needs replacing. A sagging mattress puts pressure on your spine and ligaments, and this will keep you awake.

- Do not go to bed hungry. Hunger pangs keep you awake.

- Leave an hour or two between eating and going to bed.

- Give up smoking if you are having sleeping problems. Smoking makes your blood vessels constrict, which leads to tension.

- Do not work in bed; it prevents you from switching off.

- Have a warm herbal tea to send you off to sleep. Half an hour before bedtime, the following teas help: balm, basil, chamomile, lettuce, marjoram, sage, sweet cicely, valerian or vervain.

- Gently massage the tops of your shoulders and the back of your neck to ease out tension which may have accumulated during the day.

- Play some soothing music, ideally something melodious but monotonous.

- A warm (not hot) bath or shower before bedtime has a very soothing effect. Add three drops each of juniper and marjoram essential oils to heighten the soporific effect.

- Good sex is one of the best remedies!

- Avoid coffee, tea and all fizzy drinks as they contain caffeine, tannin, sugar and frequently artificial colourings, flavourings and preservatives.

- Ideally avoid discussing domestic or work problems in bed. Have these talks earlier in the evening.

- Sleep with a fairly flat pillow and choose feather rather than a synthetic foam one. Feathers allow you to mould your head and neck into the pillow comfortably, whereas synthetics do not give enough.

SILENCE

If you are looking for perfect silence, you will find it with a flotation tank. The flotation principle was developed during the Second World War by Dr John C. Lilly as a means of researching the mental and physical effects of weightlessness and the removal of external stimuli. Further work by American and Australian scientists showed that regular floating has long-term benefits such as reducing blood-pressure, lowering physical stress levels, relieving muscle fatigue and alleviating pain, as well as promoting overall well-being.

MEDITATING

Meditation is an ancient art which has made its way into Western society and has become very popular in the last few decades. From visualization to transcendental meditation, a range of different techniques is being offered these days, either as workshops and evening courses if you want to be taught in a group with a teacher, or through a wide range of books if you prefer to learn on your own. Many people have tried a meditation technique at one stage in their lives, but few use it regularly.

Our lives can be greatly cluttered up by outside stimuli of all denominations. We are surrounded by noise pollution and an abundance of visual and sensory information in the form of newspapers, magazines, television and advertisements in the streets around us. All of these add to the stresses and strains of running a family home, bringing up children or putting in a good day's work, and some of us have to cope with all these things at the same time. A lot of energy is needed to deal with this multitude of stimuli every day, even when things run smoothly. But what if problems occur and life throws a spanner in the works? Maybe someone close to you contracts a serious illness, maybe you yourself lose your job or, just as stressful, you are promoted and feel out of your depth? Any change in life, be it positive or negative, makes further demands on our energy resources, and it is often in times of great stress that we turn to new ways of living, at least until the pressures have subsided and we feel back in control again.

MEDITATION THROUGH CONCENTRATION

We waste a lot of energy by not concentrating on what we are doing; we deal with one thing while we are thinking about a variety of other matters which we will have to attend to later. In this way we are not really absorbed in what we are doing, and this means we make mistakes more easily, we remember less well what we have just done and we sometimes even get annoyed at ourselves for being scatterbrained and inefficient.

Meditation can be a great help to become centred again and to escape the tyranny of the endless flow of mind chatter. Research shows that as you gradually achieve greater inner stillness, your heartbeat slows down, less oxygen is used and less carbon dioxide produced. Also, the blood lactate level drops, which is related to the physical relaxation during meditation – when you are anxious or tense, lactate levels are high. In addition, brain waves slow down from the beta rhythms found in the awake state to the longer alpha rhythms which promote healing in the body because all the muscles and organs are now in a harmonious state where cell repair work can take place unencumbered.

There are four so-called paths of meditation. One way is to approach meditation via a path which involves an intellectual understanding of reality. Here, the intellect is used to go behind the rational mind to transcend the will and directed thought.

Another path is the path through the emotions, which is probably the most widely practised, especially in connection with various religious beliefs. Also known as the devotional path, the meditation concentrates in believing in a god, surrendering completely and believing in this god's all-embracing love and support.

Then there is the physical path which is found in Hatha yoga, T'ai Chi and also in the Alexander

Technique. By practising single-minded concentration on the body and its movements, awareness is heightened to a level where it fills the consciousness to the exclusion of anything else. Finally, there is the meditative approach which requires action. You learn how to 'be' and relate to the world during the performance of a particular skill such as archery, aikido or karate in the Zen tradition, and singing or prayer in the Christian tradition.

The goal of all these various forms of meditation is to help you do just one thing at a time and to be at one with yourself. If physical movements are part of the meditation, it is best to learn from a teacher; this is why within the context of this book I will confine myself to meditations which do not necessarily include bodily exercises. In the following example you are asked to look at an object without putting into words in your mind what you see.

Exercise: Contemplation Meditation

- Choose something simple – a piece of jewellery such as a ring or a bracelet, or anything else that is simple and holds no emotional connotations, such as a button or a leaf.

- Place the object at a comfortable distance away from you and look at it. Feel free to bring it closer or further away from you as you explore it with your eyes and without words.

You will notice again and again how your mind drifts off, thinking about other things or translating what you see into words in your mind. Be aware that this is bound to happen a lot for quite a long time to come, and it is perfectly OK for these slips to happen. If they did not happen, you would not have any need to do the meditation in the first place! If you find your mind straying, notice this fact and then bring your mind gently but firmly back to the task in hand. Be aware that, initially, the aim of meditation is not so much to do it well but rather to do it consistently.

Set aside three to five minutes every day to do your contemplation meditation. Stay with your chosen object rather than changing to another one every week, as the process is easier if you stick to something familiar.

You will find that you are getting varying results from one session to the next. One day you may find it easy to concentrate, whereas the next day your mind is all over the place. This is normal and no reason to give up! Persevere and just do your best – and always remember that it is normal for your mind to run away with you, so do not give yourself a hard time over it. After all, you would not dream of sending a six-year-old to university, so why should an undisciplined mind (and we all have one) be able to stick to a straight line without veering off?

PART IV

Interacting and Exploring

You have by now read through a great deal of sensible
suggestions for helping your body function at its optimum
level; now it is time to loosen up and look at how you can
add energy by involving others and making a few old
dreams come true. Life is not all eating well and exercise
and resting; there are also the emotions to be considered.
Life becomes so much happier when you fill it with fun
and play, suspense and excitement. Interests, hobbies and
activities lift the spirit and provide you with a natural high
that no drug can ever hope to achieve. Abraham Maslow
(1908–1970), one of the leading humanist psychologists,
proposed that man can only reach self-actualization when
all the basic needs such as food, shelter, safety and sense
of belonging are fulfilled. At the top of this hierarchy
Maslow put self-actualization, which he described as the
full expression of a person's potential.

Part of self-actualization are the so-called 'peak experiences' which are characterized by feelings of ecstasy, sensations of being flooded with delight and a sense of limitless energy and fulfilment where almost anything seems possible. Occasions that are likely to spark off these peak experiences can be listening to music, getting totally involved in an enjoyable task, watching a beautiful sunset, or even ordinary everyday situations of being with people you really like. Peak experiences are immensely invigorating and are accompanied by a feeling of oneness with everything and everyone around you, when you no longer experience existential anxieties about death, separation and the problem of choice.

The rest of this book examines ways which can help you to release your psychological energies, those feelings of happiness and joy which lift your mood and motivate you to progress forward and become the best person you can possibly be. Mobilizing psychological energies also has beneficial effects on the body – because you are feeling good emotionally, your immune system works at its best and you are less likely to fall ill; or if you are ill, you will recover more quickly. Looking after your psychological energy levels is not a gimmick; it is vital for happiness and health.

IMPROVING YOUR ATTITUDE

Negative thoughts produce negative feelings, and negative feelings drain physical and psychological energy. Negative thoughts and feelings can keep you a prisoner in your own mind, with your thoughts dictating your behaviour, your actions and reactions. Your thoughts can really run away with you – before you know it they have produced vivid pictures in shades of grey and black about gloom and doom, disaster and failure, and as a consequence you feel terrible but still compulsively view these images in your head. How many people give up a personal project just because it did not work out immediately? They start a diet, and when they eat a 'forbidden' food after a few days they allow their mind to produce an image of eternal fatness with 'failure' written in block capitals all over it, and they give up. The same negative pattern can kick into gear with any other project, such as stopping smoking, learning a new skill, becoming more confident, improving relationships and learning to relax, to name just a few.

It helps if you can consider these negative thoughts as your inner protective mechanism which tries to prevent you from getting hurt by stopping you from doing something that could turn to failure if it did not work. Depending on how confident you are, this protective mechanism will vary in volume and frequency. When you are very unsure of yourself and your abilities, you may find that doubts and negative thoughts occur often and intensely, whereas if you are quite confident the negative thoughts are usually quieter and can be overridden more easily.

The problem with negative thoughts is that they may mean well in their attempt to protect you, but they are also stopping you from doing new things and thereby becoming a stronger person. This is why it is worth your while working on putting these thoughts into their appropriate place. You need a warning from your mind when something truly dangerous is about to happen, but not on any other occasion. Here are a few mental strategies to combat negative thinking.

STRATEGY 1

Whenever you find yourself caught up in a mental 'disaster orgy', turn the volume down! These negative thoughts are only thoughts after all, and thoughts can be manipulated and changed. They are like a film in the cinema – gripping stuff, but not real. So when those negative thoughts and images come flooding in, treat them like a radio or television programme which you do not want to listen to or watch. Imagine you are turning the volume down or switching channels to a more positive station.

Strategy 2

If you find it difficult to relate to Strategy 1, there is another way of dealing with negative thoughts to prevent them from defeating you.

When you catch yourself starting on a negative thought track, continue it voluntarily and make it purposely into an extreme tale. Say you have just started to worry about going back to work after years of having been away, perhaps because you have been looking after children until now. As your mind is now dutifully beginning to recite all the things that can go wrong, take over voluntarily and spin the most amazing yarn you can think of, such as 'Because I have been away so long, I have lost all my abilities, and I will get absolutely everything wrong on the first day, and nobody will help me, and they will all think how stupid I am, and then my boss will get really fed up with me and then she will throw me out, and then my entire family will think I am an idiot, and nobody will want to speak to me' and so on.

You will notice that, provided you make your tale

extreme enough, it either makes you laugh because the scenario is just too unreal, or you get bored with it after a while because it is so melodramatic.

STRATEGY 3

A very quick way of dismissing negative thoughts is to allow them to run to the end and just add, 'Yes, yes, OK, but I am going to do what I want to do anyway!' It is a bit like speaking to a nosy busybody who is getting on your nerves and whose verbal meanderings you cannot really take seriously.

LEARNING NEW THINGS

While you are still young and have to attend school, you remain in a constant learning process. Learning is your 'job' as a child. Some things you learn automatically in infancy, such as walking and speaking; other things you learn by being taught by parents or teachers. It is a shame that this main learning phase of your life coincides with other interesting developments such as discovering the world around you and other people, and becoming fascinated by and pursuing your first romantic relationship which, as most of us know, has a decidedly adverse effect on academic efforts . . .

When faced with a learning environment every day, you cannot help but take in some of the things that are being taught; if a subject is presented in an interesting way and you make special efforts to learn, the results can be very satisfying. If you think back to your school days, you may remember a favourite subject that you got really involved in and grew to be good at, and then you will also remember the delight of getting good marks for it.

But as we get older and there is no longer any pressure to learn, we can forget what a rewarding experience learning is. Once you are in a job, whether this is in the home, bringing up children or outside it at an office, factory, etc., you are usually not required to learn much any more, unless it is to do directly with your job. Not only that, you often do not have the time to study, even if you wanted to. The main bulk of your time is taken up with earning money, and on the weekends you use your time to recover from work.

Yet there *is* usually time for other things. Think about how many hours you spend watching television per week – if you deducted only three hours from your TV time, you could join a class or do a course to learn something new.

Remember that you can choose any area of learning. You do not have to choose anything 'useful' at all; as long as it is a learning process, it qualifies. The main criterion should be that it is fun for you to do and that it gives you satisfaction. You do not need to compete, you do not have to become good at it (although it helps if you do); all you need to start off with is an interest, and you will have found one of the most satisfying ways to pass your leisure time. Learning enhances your confidence, keeps your brain working, and often gives you a wider circle of friends and opens up new avenues for the future. For some people, learning also provides them with better career prospects or even a new professional direction.

HANDLING NEGATIVE ENERGIES COMING FROM OTHERS

It is all well and good if you are happy and in a good mood, but sometimes other people can upset the apple-cart. When you have to spend time with one or several people who are bad-tempered, critical, unhelpful or moaning a lot, it can have a very negative effect on your energy levels. If you know someone who is exuding negativity, you will know how drained you feel at the end of only one hour with him or her. This can become a real problem if you need to spend a lot of time with this person, if for example he or she is a work colleague or a family member.

The way you go about dealing with negative people will depend on your personality and also on how close you are to them. Whatever the circumstances, it is important that you do something about being exposed to these negative vibes. This has nothing to do with you being nasty or intolerant, but we all have a certain pain threshold when it comes to being in contact with a moaner or a gossip-monger, and you need to respect that threshold, otherwise you are being dragged down, and that certainly does not help anybody.

Here are a few ways in which you can handle negative people if you are not one of the happy few who can genuinely ignore them.

SMILE AND BE FRIENDLY

This method can work really well for occasions where you have professional dealings with someone who is grumpy. It does not always work, but it is worth a try. Do not be smarmy, do not flatter, do not try and jolly the person up – just smile and speak in a friendly voice.

HUMOUR

If you have it in you, humour can really save a situation, provided you are confident enough not to take someone else's bad mood personally. It is essential that any joke you make does not insult the other person because that would, understandably, only backfire and put him or her in an even worse mood.

TALKING IT OVER

If there is any chance of addressing the problem, seize the opportunity with both hands. This is by far the best method for changing things for the better. Do not forget that other people may not even know that their behaviour bothers you. They cannot read your thoughts, so do not pretend you do not mind their grumpiness if you really do not like it. We humans tend to operate on the principle that we can continue doing things our way as long as no one objects or complains about it.

Putting a Stop to It

If you have talked things through several times without getting them to stop the annoying or exhausting behaviour, you can ask them to stop it – 'Please don't tell me any more gossip, I don't like it;' 'Stop complaining, it gets on my nerves;' 'Stop criticizing me all the time, it doesn't do anything to help me, on the contrary.' Keep it short and do not be side-tracked by lengthy explanations as to why you should not mind it and how they are only speaking the truth and how life is not all that wonderful for them. It does not matter what their reasons are, just repeat that you want them to stop their moaning or criticizing or gossiping until they do; otherwise leave the room every time they do it.

Staying Away

This is the simplest solution and a good way of dealing with people you are not close to and not dependent on. A gossipy neighbour or morose work colleague who sits in another office can usually be avoided quite easily. Staying away is also the ultimate solution if you have tried any of the other methods without success. With people who are close to you, use it only as a last resort. Try everything else first, and only when they refuse to talk to you about the problem, tell them that you would rather stay away. That way, at least they know why you do not get together with them any more.

There is one category of people whom I think of as 'poisonous' and whom you should stay away from at all cost, no matter how close they are to you – because they can destroy you, knowingly or unknowingly. These are people who are violent, either physically or verbally, who will give you to understand that you cannot do anything right, no matter how hard you try, and that you will never be good enough. These people are not good enough for you, even if they are your parents, lovers or 'closest' friends. Associate with them at your peril.

ENERGY THROUGH CONVERSATION
WITH OTHERS

If other people can deplete you of energy when they are negative, those who are happy within themselves can have a very positive effect on your well-being. We derive a great deal of self-esteem through the way other people relate to us, and that is one reason to seek out the company of those we get on well with. Harmonious relationships have a soothing effect on the psyche and a healing effect on the body.

Ideally, you have a close relationship with a partner, but even without that you can derive immense benefits from being with friends and colleagues and having good conversations with them. Speaking with others is an important part of every day, and if you are on your own you need to make extra efforts to speak to others. If we do not speak to others, we live in our heads, and as our mind is such an undisciplined part of us, it tends to go walkabout, producing its own images which are not always very realistic. When you are on your own with little social contact, any shadows in your mind become greater. If you have a tendency to find fault with yourself, you will have a lot of time to dwell on your shortcomings, without considering your positive points as well, whereas when you speak to others and have a conversation with them, the mere fact that someone else takes the time to talk to you implies that you are essentially OK and worth speaking to. This does not even have to imply that the other person actively points out anything positive about you; the mere act of conversing establishes that you have value.

There is often a great discrepancy between how we see ourselves and how other people see us, and most of us think of ourselves in worse terms than other people do. We usually consider ourselves less capable and less interesting, whereas others might have a much more favourable opinion of us. Getting amicable feedback from others can redress the balance, especially if they point out examples where you proved your intelligence, prowess or good sense.

Also, exchanging thoughts and opinions with someone else helps you re-evaluate where you stand in life. Have you ever discussed relationships with a friend whose parents interfere in everything she does? It makes you appreciate your own parents more if they are helpful but let you get on with your life. Whereas beforehand, you may have taken them a bit for granted, the contrast that emerges through the conversation highlights your good fortune. Or, if matters are reversed, your friend doing really well in her new job and climbing the career ladder while you feel you are floundering a bit, this may serve as an incentive for you to ask for that long-overdue pay rise or promotion.

If you are not sure whether you talk too much, ask your friends to give you an honest opinion. It is worth knowing how you are doing in this respect; better balancing of talking and listening can actually enhance relationships.

NEW VENTURES AND CHANGES

If your lack of energy comes from a sense of boredom, from having lived the same sort of life for too long, introducing a change can be very effective. There is no need to wait for your midlife crisis to come along; you can choose to introduce a surprise element into your life at any stage, and the change does not necessarily have to be a profound one to liven things up.

When the humdrum of sleep-eat-work-eat-work-sleep begins to get to you and you start feeling listless and dissatisfied, it is time to take action. Redecorate your place, change around the furniture, sleep in the living room instead of the bedroom, ring up people you have lost touch with, go out in the evening to a club, the cinema or theatre, book yourself on an abseiling course for the weekend or cycle to work instead of taking the car. Swap domestic chores with your partner (very interesting if you have been repairing the car and your partner has been doing the cooking!), stay in a tent in the garden with your children overnight, set your alarm clock for 5 a.m. and watch the sunrise, learn to bake a cake or write a story. The possibilities are endless.

Sometimes circumstances can help you along the way. Having been ill, having lost your job or having split up with your partner can all open the way to new ventures. When life has already forced a change on you, you might as well make some further changes; things are already upside down, so any further upheavals that you have control over and make happen yourself will not do you any harm. Understandably, most people will not want to disturb the security of their present life by looking for a better job or taking time off to travel, but when fate intervenes, it can be much easier to change other aspects of your life that may have been getting you down.

WHEN CAN IT BE DONE?

Is there a natural gap coming up anyway or do you have to create a space? Which is the best point in time for you and anybody who is dependent on you?

Alone or with Others?

Is your venture dependent on others' participation? Find out who might want to join you and who could support you even if they did not participate.

Get Wise

Collect as much information about what you want to do. Ideally, speak to people who have successfully done the same thing already. Get advice from people who have first-hand experience. This is not the time to listen to well-meaning and worried relatives.

Commit Yourself

Get the show on the road! Make that phone call, put down the deposit, make the booking. It is normal to feel nervous, but as long as you have prepared yourself well, you have done what it take to make the enterprise as safe as possible.

DOING SOMETHING YOU HAVE ALWAYS WANTED TO DO

We all have dreams about what we would like to do in life, dreams about achievements we would like to accomplish, possessions we would like to call our own, feats we would like to dare and places we want to see. These are sometimes extravagant fantasies, but often they can be realized, it is only that we have not had time so far, or it has never been the right time, or we think that we 'should not' want to crave what we do not have. And yet the dreams persist, and we carry them with us, if only at the back of our minds, often taking them to the grave. What a shame! So many of these dreams can become reality, if only we put our minds to it and persevere.

We need to fulfil for ourselves some of our heart's desires to be happy in life. For some people, this can be their job or their children, for others it is travelling or owning a particular car, for others still it is to acquire a particular skill such as singing or drawing.

Exercise 1

Think about your own dreams. What is it you have always wanted to do in life? Let your mind wander all the way back over the years that have passed. What were your dreams when you were a child? What are your dreams now? Make a list.

Now look through your list and decide which things you could do if you really wanted to.

Exercise 2

Pick out the one dream from the 'feasible' list that is closest to your heart. Write down what steps you would have to take to achieve your dream.

A dream is something that is exclusively yours and reflects what you are all about. It is perfectly legitimate to want to fulfil a dream, as long as no one else is adversely affected by it. Pursuing a dream can keep you going when other things do not work out so well in your life; it can be an anchor that allows you to stay centred in turbulent times.

When your physical energy levels are low you may initially be excluded from strenuous activities, and some of your dreams will have to wait until you are stronger – but there are still possibilities. A dream can have a high motivation factor in getting you better, simply because you have something desirable to strive for. Concentrating on your dreams not only keeps the nightmares at bay, it also helps create psychological energy which in turn helps build up your physical energy.

PART V

Creativity and Play

In the chapters that follow, you will find a collection of different suggested activities and pastimes which all have one thing in common – they are creative. Even if you have gone through life so far thinking that you have not one creative bone in your body, you will be able to find something that you can do and that you can learn to do well. The projects and exercises you will read about are meant as tasters; no great financial outlay is involved nor do you have to exert yourself at the expense of your possibly low energy levels.

The purpose of these activities is to give you something pleasurable to think about, something to focus on which gives you a positive buzz, something that you can get enthusiastic about and look forward to.

You may have already tried some of these ideas and even enjoyed them, but then perhaps something happened and you could no longer pursue your interest. In that case, let this section help you rediscover how uplifting and rewarding it is to do anything that is creative.

PLAYING

Once we have become so-called grown-ups, we tend to spend most of our time with activities that are sensible, necessary or good for us. We go to work to earn money so we can pay the bills, buy furniture, pay the rent or the mortgage, send the children to a good school, feed and clothe them and us, and hopefully have something left at the end of the month to allow us to save up for a holiday or for the occasional special evening out. We exercise our bodies by going to fitness classes, we play tennis or squash regularly because we know the body needs to be challenged and stretched to function properly and because it helps reduce stress and generally makes us feel better. We watch what we are eating, we go to the dentist for check-ups, we have our car serviced regularly and we do a hundred thousand other things in life because they are the right thing to do. So what about activities that are optional, that do not serve any particular purpose, that are neither education nor sensible?

The further we move away from childhood, the more likely it is that we feel we need a good reason to do anything that is fun. Inhibitions have started piling up over the years, and playfulness becomes a decided handicap in the race for a serious career. In order to be recognized as a potential candidate for higher office, you have to fit a certain image, wear certain clothes and behave in a certain manner. The moment you step out of that prescribed framework, you are in trouble. Unless, of course, you are rich or famous (preferably both), in which case any deviation from the norm becomes an eccentricity which is smiled upon rather than frowned at.

What happens to our ability to play once we have grown up? What happens to the helpless giggles, the pillow fights, the looking at clouds and describing their shape, and running around, yelling at the top of your voice that so-and-so is 'it'? Sometimes, you see couples in love doing some of these things; sometimes you see adults who play with their children's toys, but what about the rest of us?

Play frees us from a lot of conventions. It is like a childhood mini-meditation where you get totally wrapped up in doing one particular thing, such as balancing on a wall or playing house.

ACTING

Acting is quite similar to playing in that it provides you with an opportunity to 'make-believe', just as little children do quite naturally when they play house or school. They slip into another persona, and they are usually very competent at it. Have you ever watched your children playing and wondered whether you really sound the way they are imitating you or your partner? (Chances are you do!)

In later life, quite a few barriers come up against slipping into another role, even though we do it often anyway, just in a less drastic way. Think about it – you may find that there is a slight difference in the way you speak and behave towards your boss compared to the way you communicate with your colleagues; you are probably more formal with the one than with the others. However, play-acting requires much greater changes, and the main objection people have to it is that they would feel 'silly' doing it. It seems such a childish thing to do, and once you are involved in a career, you seem light-years away from childhood where you could improvise quite happily and naturally.

If you go for an Introduction to Acting course, you may find that the main focus is on improvisation. But before you get really started, the teacher will help you loosen up and relax, maybe by doing some physical exercises and jumping around, so that by the time you get started on a little scenario, everyone is laughing and talking to one another and it becomes much easier to adopt a role which may be very different from your everyday person. Your teacher will know about the inhibitions students bring along with them, and is trained to ease you all gently into the process of overcoming them.

Scenarios your group may be asked to act out could include little everyday situations such as being in the supermarket, or a day in the countryside, or being a teenager who has had a bit too much to drink. Usually, the students have a little discussion to decide how to tackle the scenario before having a go. Others might be watching and afterwards commenting on your body language, for example, or adding some suggestions. Do not be too worried about criticism, though – they will have to act next, and you will be watching!

Classes might also include voice control and correct breathing to produce resonance, and you may also be taught about movement and in this context explore various ways of walking – slowly, frantically, hesitantly, in a nonchalant manner, and so on.

After the introductory lessons, you can go on to a two-year course if you want to take acting further. You will now begin to focus quite strongly on body work and voice work, and you will also go much more into emotions, which means you will have to refer back to yourself and past experiences.

Whether you go for a one-day workshop or a course of lessons, remember that this is meant to be fun. There is no point in going about these classes in a competitive way; just do your best, and that is fine. What counts is that you learn and explore in more detail who you are and what different parts you consist of. As you get better acquainted with yourself, you will feel more confident and will find that your acting class helps to bring out those sides of your personality which can help you to deal best with the various life situations you find yourself in.

LAUGHING

You may consider laughter a pleasant but trivial matter that has no bearing on your energy levels. The fact is that it does, and research has now proved that laughing and smiling have considerable beneficial effects on your body. As you laugh, your belly area and diaphragm are exercised – remember how people who laugh hard often hold a hand on their belly? Hence the expression 'a belly laugh'. This abdominal movement deepens breathing, which in turn gets more oxygen into your body and makes circulation more effective. It has been found that laughing expands blood vessels, which helps to speed up tissue healing. On top of all these advantages, laughing also stimulates the production of endorphins (the body's natural pain-killers) as well as helping the lymphatic system rid the body of wastes, burning off fat and generally relaxing all the muscles.

On the emotional side, laughing is indeed the best medicine as it is impossible to laugh heartily and at the same time be anxious, tense or depressed. When you have a good laugh about something you have seen on TV or read in a book, your whole body is in motion and, just for a few minutes, you can forget yourself and any worries you might have. Watch people when they are overcome by laughter – they loll around in their chair helplessly, they bend forward, they throw their heads back in mirth and do not care what they look like at that moment. It is a lovely way of losing control for a moment or two!

In many ways, laughter is the friendly cousin of anger, sadness and hate. It acts as a safety valve which will allow negative emotions to escape in a positive way so that they cannot do any physical damage to you. It is by now well established that people who are happy within themselves enjoy better health and are less prone to disease than those who bear grudges and like neither themselves nor others.

Laughter is also a good remedy against exaggerated self-importance and taking yourself and life too seriously. Rather than being annoyed at yourself for having made a mistake, why not simply laugh about it? Rather than living as if the world revolved around you, why not get life into a more realistic perspective and laugh about yourself sometimes? Both a self-critical attitude and an inflated ego ultimately stem from the same root – fear, and fear can often be combated with a sense of humour.

You may think that this is all well and good, but that there are not that many opportunities in your life for bursting into spontaneous laughter. And anyway, with your financial worries and complicated love life, not to speak of your unsatisfactory job, it is no wonder that your sense of humour has finally abandoned you . . .

And yet, it is in precisely these situations that humour and laughter can be your saving grace. Laughter blows the cobwebs away and allows you to think more clearly about what you need to do next. If you are dissatisfied with certain aspects of your life, you can become over-anxious about them, and this state of mind is not conducive to finding a solution to your problems. Shake that tension loose with a good laugh, get your muscles and organs working again and you will be able to function better.

DRAWING AND PAINTING

Reproducing shapes, people, landscapes and colours on a piece of canvas or paper is a skill, and most of us have been exposed to some instruction in this skill to a greater or lesser extent at school. Children like to paint and draw, and they do so with admirable zest and dedication, with practically no inhibitions. It is usually adults who feel embarrassed because they cannot make out what their child's drawing of a red tangle with a blue blob and three sticks that look like legs could be, only to be told that *of course*, it is a car!

Much of the natural joy of drawing and painting is lost later on when these little works of art are being judged and marked and comparisons are drawn with other pupils who receive better marks for their efforts. This is when you get labelled untalented, and we often hang on to our labels at later stages of life without questioning them or checking them against reality every once in a while. So do not be put off if you did not get top marks at school for your drawings; if you are fascinated by the subject, give it another go. Things change as we get older, and talents can sometimes come to light only when conditions around you have changed for the better. Having a more patient teacher or working in a smaller group with more help available for each individual can make all the difference.

Just for a moment, leave aside your preconceived ideas of what you can and cannot do when it comes to drawing and painting. Also, leave behind the concept that only one particular way of using a pencil, watercolours or oil paints is the correct one, and forget that everything you draw or paint has to look as it looks in reality. What would have happened to Picasso if he had allowed himself to be limited by the fact that people have only two eyes (well, at least all the ones I know have two) when he started painting them with three?

Exercise 1: Drawing

- Think of a simple object in your home, something without frills or details, such as a bowl or a candlestick, a lampshade or a vase.

- Without looking at the object, sketch a rough outline on your pad of what you remember about it.

- Now look at the object and take your time to scrutinize it very carefully. Which aspects of it had you remembered correctly? Which aspects had you forgotten about?

- Now compare the object with what you have drawn.

- With the object in front of you, draw it again, this time checking back carefully when you have drawn each line, to see whether it corresponds with what is before you.

Exercise 2: Painting Colours

- Use one sheet of paper to try out all the colours that you have available.

- Experiment with darker and lighter shades of one colour. See what variations are possible.

- Mix colours by letting them overlap, to see what effects you get.

Exercise 3: Expressing Ideas with Colours

- Think of an idea or concept that you find interesting. This could be 'nature', 'light', 'love' or anything else that appeals to you.

- Find shapes and colours that express this idea for you. The shapes can be abstract or natural ones. For example, some people associate 'comfort' with a cozily lit room and an armchair on a winter evening, so that they might choose yellow and orange hues and maybe the outline of a comfortable chair.

DANCING

'I can't dance!'

Well, you do not have to be good at it to enjoy it. There are an amazing array of types and styles to choose from, from classical ballet and ballroom dancing to modern dance, rock 'n' roll, barn dances, sequence dancing and creative dance. For some types of dancing you obviously require a partner, but for the purposes of this book I would like to concentrate on those styles that can be carried out by one person on his or her own.

Even though dancing constitutes physical exercise, the two are not the same. Exercise is mostly a very functional activity that aims at increasing your stamina and muscle tone and training certain muscle groups to make them stronger or more supple. The goal, in other words, is to work out and limber up the body, which will then lead to stress reduction and an increase in energy. With dancing there is an added dimension to the physical movement side, and this is creativity and the expression of feelings. Movement is now no longer a mere vehicle to exercise certain muscles, but instead becomes a story in itself. When you hear a piece of music, the melody, the intonation, the changes in rhythm or tempo all convey moods and sentiments that you strive to capture with your body movements and which you can express with the help of your feet, your arms and hands, the tilt of your head, your facial expression and the way you set your steps.

If your energy levels are fair, you can join an evening or weekend class and get instructed in the ways and possibilities of expressing yourself in dance, but you can also do a lot by yourself. Remember that there is no right or wrong about the way you choose to translate music into movement. Also, if you feel a bit inhibited about being creative amid a group of people, you might find that the following exercises, which you can do in private, give you more confidence so that you feel able to join a class later.

The following exercises are very therapeutic if you are someone who finds it difficult to speak about feelings, as they represent ways of expressing emotions without words and thereby releasing them.

Exercise 1: Dancing with Your Hands

- Choose a favourite piece of music you have on record, tape or CD.

- Play it through once and listen with closed eyes. Notice which emotions it addresses in you and whether the music tells you a story. Sometimes, the title of the piece can give a theme for a story.

- Play the piece again, and this time, only use your hands and arms to express the perceived emotions. (If you feel very low in energy, you can do this while you are lying down, just using your hands, or sitting up, using either your hands only or your hands and lower arms.)

Exercise 2: Dancing with Your Body and Hands

- Stand upright with your feet slightly apart.

- Let the music play for a few minutes to capture the mood.

- Begin moving your hands and arms in response to the music.

- Now add any body movements that will allow you to keep your feet where they are. Move your head to allow your eyes to follow one hand in its movement, then the other. Stretch your body upwards to follow a raised arm. Explore sideways movements and forward tilts as well, leaning forwards or backwards. Use your knees. If you wrap your arms around you when the music is melancholy, crouch down deeper, folding up your knees. Use the options of height and depth that your knees give you.

Exercise 3: Dancing with Your Whole Body

- Listen to the music to capture the mood.

- Begin to walk through the room in a way that is appropriate to the music. Try a swaggering step with a waltz, a dainty little quick-step on tip toes with ballet music, wide strides with weighty music, and so on.

- Get your arms in line with your stride. Have your arms akimbo with the swagger, swing them in an exaggerated way when you stride out or make delicate hand movements while your arms stay by your side when you walk on tip toes.

- Now use any other parts of your body, adding, for example, head, shoulder or hip movements.

- If you want to be particularly creative, dance a little story to the music and integrate objects in the room, carrying a vase from the window to the table while using your chosen step, or designing your steps around a chair, for example.

WRITING

When I am in the process of writing a new book, I will often go out to a park or drive even further out of London and sit in the grounds of a country house, leaning against a tree and enjoying the air and sunshine while at the same time getting some writing done. In winter, my friend Ursula and I will sometimes go off to a hotel with a comfortable lounge where we can sit around a cozy fire and write. It is amazing what you can get done when there is no telephone and no fridge there to distract you . . .

I derive a lot of pleasure from writing, and whenever I have finished yet another chapter I feel I have achieved something and feel good within myself. And, as you can imagine, once the book is finished, it is celebration time! But you do not have to write professionally to get an energy-boost out of writing. If you have ever kept a diary, you will know what a gratifying and often therapeutic process writing is. It is also interesting to see what happens when a thought in your head is transcribed onto paper; the thought or memory reads quite differently once it is written down, compared to the way it feels when it is merely in your head. One of my clients started writing down childhood memories, which unfortunately were not of the pleasant kind. When she read through what she had written, she said it was nearly as if it had not happened to her but to someone else, and at the same time it sounded doubly tragic because she could now look at it from the outside and realize the enormity of those past events.

However, on a lighter note, if you would like to write but are not sure what to write about, why not pick a happy memory from the past? You could try your hand at something funny or unusual that happened once, and write it down as you remember it. Or you could just use a fragmented memory as a foundation for a little story, filling in any gaps with your imagination, or spinning out a minor event into a longer story.

Diaries are a wonderful way of preserving the present for the future, and you may want to write little accounts of what your children are getting up to while they are still little, or, if you have a pet such as a cat, what is happening with them. Again, any of these little events could lend themselves as themes for a short story.

Those of you who want to get into more serious writing might want to think of taking an evening class in creative writing or even going on a writing holiday. These holidays last between three and six days. On the longer courses you might be asked to choose two topics from among a wide range of themes, which might include fiction, non-fiction, writing for radio, writing for television, journalism or historical novels, for example. Your mornings might be taken up with lessons in your two chosen subject areas. In the afternoons, you could have talks by experienced writers, agents or publishers on their specific area of expertise, and after dinner a well-known writer might give a fun talk which is designed to be entertaining as well as informative. Between your afternoon lecture and dinner, there are usually a few hours to work on any 'homework' that you might want to hand in to your teacher for assessment. This would normally only be a short piece of work, around 200 words or so; you might be asked for example to describe the heroine of your own (theoretical) romantic novel, or to outline a scene in a thriller. The teacher would then mark your work and you would have an opportunity to discuss it with him or her.

ENERGY THROUGH CREATIVE VISUALIZATION

If you have tried any of the relaxation exercises, you will have already experienced what visualization is like. It is a bit like watching television in your head while your eyes are closed, with mental pictures that are not necessarily the same as 'normal' visual images. Many people are discouraged because they do not get the same clarity of image in their mind as they do with their eyes, but in fact it does not matter how sharp your mental picture is; even when you have only an idea in your mind about an image, that will be enough to work with.

There is a test I always give my clients who doubt their ability to visualize – I ask them to describe their living room at home to me. They will then usually move their head and point their hands towards imagined objects such as a sofa or table while they describe their living room, obviously accessing remembered pictures of the room in their mind. If I ask any more specific questions, for example what the pattern of the curtains is like, they even narrow their eyes while staring at a spot on the wall so they can 'see' the pattern more clearly – a pattern that is only in their mind! If you can describe a room in your home to someone else, you have all the imaginative powers necessary to do the exercise I am going to detail in a moment.

Spending time imagining something in your mind has a measurable effect on your body. You will have experienced the truth of this statement many times yourself – when you are mulling over a problem you encountered at work during the day, you tense up, your mind is alert and you cannot go to sleep. When you think about a grudge you hold against someone, your body tightens up and you feel physically uncomfortable, your shoulders stiff and your jaw muscles clenched. Your thoughts translate directly into physical responses, down to the most minute level of blood chemistry. You can actually measure the body's reaction to different types of thought. In my positive-thinking workshops I can use a skin tension biofeedback meter – you can tell by the direction in which the needle on the display is going whether the person is thinking about something pleasant or unpleasant.

Visualization Exercise
This energy visualization exercise is best done in conjunction with a relaxation exercise (*see page 88*); being relaxed beforehand heightens the effect of the visualization considerably.

- Make yourself comfortable. Sit or lie down and loosen any tight clothing. Close your eyes and go through a brief relaxation routine (progressive relaxation is best here, *see page 90*)

- Imagine you can look inside your body and think about energy and where your physical energy is located in your body. Do you imagine it to be in your belly area, in your heart area or anywhere else within yourself?

- Now imagine what this energy looks like. Is it like a liquid? What colour is it and what texture? Is it like a light? Is it warm or cool?

- How does the energy move through your body? Does it flow through the veins or does it radiate from a particular point in your body to all the other areas? Do you imagine that the energy causes any physical sensations like tingling or warmth when it starts flowing?

- While you are concentrating on imagining your energy system, imagine a mechanism that makes it flow freely. Maybe a symbolic 'stress tap' in your mind needs to be shut down; or perhaps there is some sort of sluice gate that needs to be opened up (hydraulically of course; even if it is only in your mind, you do not want to exhaust yourself!); or you might simply imagine a door opening.

- Operate this energy-releasing mechanism now and watch in your mind's eye how the energy starts flowing through your system, with all the pleasant physical associations which come to mind when you think about increased energy. It does not matter whether you can actually feel these sensations in reality; even if you just pretend, your body will still respond favourably.

This energy visualization exercise should be done meticulously and at length the first few times until you have a clearly established idea of how your energy works and what it looks like. The more thoroughly you do it the more likely you are to get a real physical sensation of well-being and strength. Always strive to get that positive feeling, either just in your imagination or in reality. As soon as you have that pleasant feeling, try to hold it for a moment, and then stop and end the exercise. Some people get really efficient and can produce the energy feeling in a couple of minutes. In this context, it is a great advantage if you can concentrate well. You can further this ability with simple meditation (*see page 97*).

MUSIC

There are various ways in which music can create energy in you, and each way is equally valuable; it is really a matter of personal preference and also partly a question of money as to which one is for you.

Playing an instrument will, of course, mean having to buy the instrument itself unless you already own it. Some people have played as an adolescent and later given up, but still have the piano, guitar, violin or wind instrument in their possession. If you are a beginner, a second cost factor is that you might want lessons to make sure you learn the basic techniques correctly. However, you may have reached a fairly high standard as a youngster and now only want to play a little for pleasure, and that is obviously gratifying as well, especially when you play pieces of music that have a sentimental connotation because they remind you of a certain era in your life. Striving to play your first little piece well and accomplishing this for the first time is one of the most rewarding experiences (not just for you but also for your neighbours, who have to listen to you practising!) I remember one of my proudest moments was on music day at school when I played a simple little piece by Dvorak on the piano in front of an audience of parents, teachers and students; even though I was a bit nervous I not only played it without a mistake but was also able to express the atmosphere of the piece well enough for our old sourpuss of a music teacher to give me some praise for it afterwards.

Another way of making music is of course with your voice. If the breathing exercises in Part Two of this book do not appeal to you or if you get bored with them after a while, singing is an excellent alternative. Tune your radio to a station with pop songs or golden oldies, or put on a tape with songs you like and sing along! Use your own phrasing or imitate the singer's. Bathrooms are great places for singing; they give your voice resonance and it all sounds better. If you sing along to two or three pop songs a day, you will have done nearly ten minutes of expanding your lungs and challenging your breathing in a pleasant and diverting way.

If you know that you can sing in tune, you might also want to consider singing with a choir. This is like singing in the bathroom, only better! As you have a group of people singing harmonies, the denseness of the sound you are all creating weaves a magic carpet of music; it is like being immersed in a sea of music made by voices, and yours is one of them.

GARDENING

If you really feel that you have no creative talents, why not let nature be creative for you? It does not even matter whether you have a garden or not – there are tubs, urns, troughs, pots, window boxes and hanging baskets that can go onto sills outside or on the front steps, or herbs and flower bulbs that you can grow indoors. This does not have to be complicated, as you can see when you look in the gardening section of your local bookshop. There are a great many books on the subject, and you are best off choosing something simple to start with. I found a delightful book by Ursula Markham called *The Children's Gardening Book*, which explains in simple terms what you can plant and grow from bulbs or seeds, with or without a garden. Even though it is a children's book, it is equally suitable for adults who want to learn about basic gardening. In the context of energizing, I would like to look at just a few simple ideas for indoors that anyone can follow, even without a garden.

Apart from the visual pleasure gardening will give you, you will also notice an increase in energy brought about by handling the earth and living things like plants, seeds and bulbs. Also, remember that during photosynthesis plants take in carbon dioxide (which is what we breathe out) and release oxygen into the air, so that plants act as mini-air cleaners in your home.

Here are a few easy gardening projects that you might like to try. The first example is an idea adapted from Ursula Markham's book.

Project 1: Watching it Happen

- You will need an avocado stone, four wooden cocktail sticks and a jam jar.

- Soak the avocado stone in water for 24 hours.

- With the pointed end of the avocado stone upwards, push the cocktail sticks a short way into the stone from all four sides so that the stone can sit in the opening of the jar without dropping in.

- Fill the jar with water to the point where the bottom of the stone just touches the water. Keep the water level at this height by topping it up when needed.

- As soon as the stone develops roots at the bottom and shoots at the top, plant it in a pot of compost and watch an attractive plant develop.

  Tip: To keep the water pure, drop a piece of charcoal (from drawing supply shops) into it.

Project 2: Spring Show in Winter

- You will need a round bowl of about 12 in/30 cm diameter, special bulb fibre or compost, five tulip bulbs and nine daffodil bulbs (if your bowl is smaller, use fewer bulbs). If using bulb fibre, your bowl will not need a hole in the bottom.

- If you are using bulb fibre, mix it with water so that it is moist but not dripping wet. Compost can be used as it is straight from the bag.

- Fill the bowl two thirds full with fibre or compost.
- Place the tulip bulbs in the centre in a circle, but make sure they do not touch one another. Now arrange the daffodil bulbs in another circle around the tulips, making sure that the daffodils do not touch either the tulips or the sides of the container.
- Fill the bowl with fibre or compost so that only the tips of the bulbs are visible.
- Store the bowl away in a dark, cool place. Check once a week whether the fibre is still damp. If it is not, add a little water.
- Once you can make out little green tips, you can bring the bowl out into the light. Avoid bright light until the shoots have turned dark green. Soon you will have a lovely show of colour. Uplifting when it is grey outside!

  Tip: If you want the flowers out for the festive season, look for 'especially prepared for Christmas' on the bulbs' packaging.

PART VI

Alternative Ways of Unblocking Energies

So far, you have been reading a selection of self-help methods that constitute ways of building up energy; if you are on a reasonably even emotional keel you will get very good results by putting just one or two of these methods into practice.

There is a lot you can do yourself to boost your energy levels, but sometimes you might need some outside help. If emotional issues are at the bottom of your low stamina it can be difficult to get to grips with the problem, simply because you are too close to it to be able to change negative feelings into positive ones. When there is a block from the past in your mind it can prevent you from progressing satisfactorily in life, and things that other people can do easily become a threat to you. You may be able to understand on a logical level that your negative feelings and reactions are exaggerated, but that is not the same thing as being able to change them.

I have added this last section to present you with a small assortment of alternative therapies that can help with both the physical and the emotional sides of fatigue. You will see that some of these therapies approach energy problems from the psychological side, others from the physical side; but invariably they will ultimately have an effect on both body and mind.

ANALYTICAL HYPNOTHERAPY

Sometimes a depletion of energy is due to one or several traumatic events from the past that have left you anxious and lacking in confidence. Such events can impair your self-esteem severely, leaving you fearful about matters that a confident person would not even think twice about. You may feel stressed out by everyday matters which others can deal with easily, and this makes you feel even more incompetent and more reluctant to tackle problems. These constant daily pressures and stresses will of course take their toll after a while; neither body nor mind can function on constant anxiety overload.

In order to release these fears and re-establish your inner equilibrium, it may sometimes be necessary to have emotional 'tidy-up' which includes looking at the root cause of the problem and working through it.

As you grow up from child to adult you are exposed to a great many influences from the people around you, and these, together with the particular personality structure you are born with, will result in certain patterns of thinking, feeling and reacting. When you are a little timid to start with and you spend a lot of time in an environment where others are anxious, you will in all likelihood view the world as a frightening place in which you have very little control over what is happening to you. You can also come to just that conclusion if you grew up in an environment where people were cruel to you – and whether this happens in an open or underhand way does not really make a difference.

The reason why hypnotherapy can help comparatively quickly is because it deals with the subconscious level of the mind, rather than with the logical level. Think about it – if you are someone who tends to rush everywhere and overworks constantly, you usually know on a conscious level that this is not doing you any good, but knowing this logically still does not let you change your behaviour, because the urge to rush about and overwork is not a logical problem but an emotional one. Something inside is driving you, but you do not know what that something is.

In order to help you change your perception so that you can feel more powerful and in control, the therapist will help you into hypnosis, a pleasant, relaxed state not unlike day-dreaming. Once in hypnosis the therapist will help you to regress through memories to the time when the problem started, and will help you to work through these memories. The root cause of your problem is in all likelihood something that you remember anyway, but you might have forgotten what an impact it had on you at the time. In my experience, it is rare that a client remembers something that comes as a great surprise to him or her.

In the light of recent cases of 'false memory syndrome', when choosing an analytical hypnotherapist it is worth finding out whether he or she believes that all emotional problems stem from some form of childhood abuse. If the hypnotherapist subscribes to this school of thought, he or she may indeed try to foist a false memory upon you of sexual abuse which never happened – so beware, and find yourself another therapist!

EQUALLY GOOD ALTERNATIVES

Neuro-Linguistic Programming (NLP)
A non-analytical approach in which you learn mental techniques to change any unwanted reactions you have to specific questions. NLP is used with and without hypnosis.

Brief Therapy

Based on Steve de Shazer's work, this approach is about discovering and expanding on existing methods that the client is already using unwittingly to combat the presenting problem.

Counselling

The therapist helps the client to explore present behaviours and find more constructive ways of dealing with difficult situations.

Gestalt Therapy

A very practical type of psychotherapy in which the client is encouraged to act out old and new behaviours and feelings. Gestalt therapy can take place in a group or on a one-to-one basis.

SHIATSU (JAPANESE MASSAGE)

Energy in our bodies flows along certain energy paths, known as meridians. Acupuncture, acupressure and Shiatsu points are located along these meridians. There are 12 of them, and each meridian is named after a physical organ, for example the Heart Meridian or Lung Meridian. But the meridians do not only relate to particular organs in the body, they are also associated with certain functional qualities, such as protection, control and planning, detoxification, and so on.

If you carry toxins or wastes in your body or if your muscles are tense as a result of negative emotions, energy can no longer flow freely because it is hindered by these blocks. During a Shiatsu session you lie on a padded mat on the floor, dressed in comfortable and preferably loose clothing. The Shiatsu practitioner will apply pressure to the various points on the meridians with his or her thumbs, fingers, palms and sometimes elbows, knees and feet to stimulate the flow of energy. When the practitioner exerts pressure on a blocked point, you may experience discomfort because the pain nerves in this part of your body are under greater strain due to the fact that the blocked muscle is pressing on surrounding blood and lymph vessels.

As various points on your body are pressed, kneaded or stretched, physical energy imbalances are corrected. At the same time, you may also experience a release of emotions on some occasions – as the physical blocks are freed up and removed, the emotions can come to the surface and be discharged. This does not always happen during the session; sometimes you will only become aware of the emotional effect later in the day or during the following week. I have also heard from some of my clients that Shiatsu brought back memories they had long forgotten.

The duration and frequency of treatment will vary from person to person, as will the total number of sessions required. After a session you can expect to feel invigorated but relaxed.

EQUALLY GOOD ALTERNATIVES

Acupuncture
Rather than using hands, acupuncturists use very fine
needles which they apply to acupuncture points. Other
methods of stimulating the points include electrical
stimuli, mechanical vibrators, heat oscillators and
magnetic oscillators.

Acupressure
Similar to Shiatsu, except that in acupressure only fingers
and thumbs are used to apply pressure. Also useful as a
self-help method.

Reflexology
Reflexology concentrates on the hands and feet only,
where certain points are stimulated that correspond to
certain organs.

HOMOEOPATHY

Homoeopathy involves the administration of therapeutics which are derived mainly from the natural world – that is, from plants, minerals and animals. The fundamental principle is that what a remedy can cause, it can also cure. This means that the remedy needs to match the totality of the patient's symptoms in order to be effective. For example, if your complaint is that you are feverish and your pupils are wide and dilated, your homoeopath would prescribe one remedy, whereas if you had a fever and your pupils were normal, you would be given a very different remedy.

Homoeopathy is tailor-made to your specific needs, and it takes into account all aspects of your personality and your life. This is why your homoeopath will spend considerable time with you before starting treatment, finding out all the details about your symptoms as a GP would, but then going into an even more detailed examination about any injuries you might have sustained and any bereavement, fright, shock or even resentement you might be harbouring. You will be asked about your basic emotional and mental characteristics, whether you are usually relaxed or highly-strung, shy or outgoing, impatient, irritable, sensitive, suspicious, whether you bottle up problems or talk about them.

With the help of this information, the homoeopath can decide on a remedy or a selection of remedies. In acute cases where the symptoms are more clear-cut, this choice can be made more easily than if you come in with a chronic condition where the illness has many facets to it and needs to be removed layer by layer.

Remedies are administered in a range of potencies, either diluted in water or alcohol, or ground up with lactose. Dr Samuel Hahnemann, the originator of homoeopathy as we know it today, discovered that if he diluted substances in this way, they became more effective than if they were used in their crude form. Homoeopathy sees illness as an imbalance that needs to be corrected from the inside to the outside; first, the inner organs like the liver, heart and lungs, then moving out to the outer organs such as the skin.

EQUALLY GOOD ALTERNATIVES

Herbalism
This therapy uses only plants or specific parts of plants to cure illness, and these are used in their undiluted form. Herbs are grown and harvested under very specific conditions to have the maximum healing effect.

Naturopathy
A naturopath helps the patient create the best environment to stimulate the body's own healing powers by prescribing homoeopathic remedies and herbs, by correcting posture and breathing and by teaching relaxation methods. During treatment, iridology is used to diagnose in the first instance (this is done by checking the appearance and condition of the patient's iris) and then later to monitor progress.

Epilogue

So now that you have read through the book and looked at the various possibilities for restoring lost energy, you have noticed how some methods appeal to you more than others; you may even have started following one or several of the suggestions.

Remember that you can tailor-make any of the methods to suit your own specific needs. For example, if you cannot do your stretching exercises every day, do them every other day. If you like the idea of meditating but not the particular exercises suggested in the book, change them until you like them better!

Creating health and energizing yourself does not require you to be perfect. Even if you do not get it right 100 per cent of the time, this is nothing to worry about. Energizing yourself is not about perfection; it is about getting started. As long as you start gently and then gradually increase your personal energy programme, you will succeed. It is very rewarding to see the first improvements, and you will find that your increased stamina will motivate you to make further positive changes. If others can do it, so can you!

Also available …

The Sleep Technique

Simple secrets for a deep, restorative night's sleep

ANTHEA COURTENAY

When was the last time you had a blissful night's sleep? Do you get into bed and just toss and turn?

Packed with simple and effective techniques, this little book shows you how to beat insomnia and wake up full of energy to last all day long.